RESOURCES FOR THE FUTURE LIBRARY COLLECTION
NATURAL RESOURCE MANAGEMENT

Volume 4

Comparisons in Resource Management

Six Notable Programs in Other Countries and Their Possible U.S. Application

Full list of titles in the set
Natural Resource Management

Volume 1: New Deal Planning
Volume 2: America's Renewable Resources
Volume 3: Land Use and the States
Volume 4: Comparisons in Resource Management
Volume 5: Perspectives on Conservation
Volume 6: The World Copper Industry
Volume 7: Conservation and Economic Efficiency

Comparisons in Resource Management

Six Notable Programs in Other Countries and Their Possible U.S. Application

Henry Jarrett

RFF PRESS
RESOURCES FOR THE FUTURE

Washington, DC • London

First published in 1961 by The Johns Hopkins University Press for Resources for the Future

This edition first published in 2011 by RFF Press, an imprint of Earthscan

Earthscan LLC, 1616 P Street, NW, Washington, DC 20036, USA
Earthscan Ltd, Dunstan House, 14a St Cross Street, London EC1N 8XA, UK
Earthscan publishes in association with the International Institute for Environment and Development

For more information on RFF Press and Earthscan publications, see www. rffpress.org and www.earthscan.co.uk or write to earthinfo@earthscan.co.uk

ISBN: 978-1-61726-051-3 (Volume 4)
ISBN: 978-1-61726-006-3 (Natural Resource Management set)
ISBN: 978-1-61726-000-1 (Resources for the Future Library Collection)

A catalogue record for this book is available from the British Library

Publisher's note

The publisher has made every effort to ensure the quality of this reprint, but points out that some imperfections in the original copies may be apparent.

At Earthscan we strive to minimize our environmental impacts and carbon footprint through reducing waste, recycling and offsetting our CO_2 emissions, including those created through publication of this book. For more details of our environmental policy, see www.earthscan.co.uk.

CONTENTS

LIST OF MAPS

EDITOR'S INTRODUCTION

FOR ALL THAT IT LOOKS ABROAD at how other countries are managing their land and water resources, this book has its roots in home soil. Each of the six cases examined here deals with a basic problem that is important in both the United States and the other country, and the final aim in each instance is to discover what useful lessons Americans can learn from the foreign experience. In a way, then, this volume is a modest venture in importing Technical Assistance into a country which in recent years has mostly been concerned with how best to export that valuable and volatile commodity. It is not intended, though, that the United

■

THE EIGHTEEN ESSAYS in this book were originally presented as public lectures in the 1961 Resources for the Future Forum on Comparative Resources Policy and Administration. All of the papers have been revised by their authors prior to publication without, however, any significant changes in substance. Maps and photographs were supplied by the authors of the essays with which they are used. Many of the maps were redrawn for reproduction by Clare O'Gorman Ford. The lectures were given in Washington during January, February, and March, 1961, in the Falk Auditorium of The Brookings Institution building where the RFF offices now are located. Irving K. Fox, vice president of Resources for the Future, took the lead in planning the series and developing arrangements for it. Other staff members who made major contributions were John E. Herbert and Francis T. Christy, Jr. Joseph L. Fisher, president of RFF, was chairman of each of the six programs. The 1961 Forum was the third such public program to be sponsored by Resources for the Future.

States be the only beneficiary of these studies in comparative resource administration. In the close-knit modern world all countries have much to learn from each other.

The questions explored range from one as specific as how to increase yields on small forest holdings to one as broad as how best to administer national parks as population becomes denser and available land scarcer. Each, however, arises from a currently significant resources problem that seems more likely to grow than to diminish and on which the American people probably will be making important decisions during the 1960's.

For the small forest holdings an urgent question, often stated but still unsolved, is how to get better management to raise timber yield on many of them. This country is going to need more timber in the years ahead; average productivity is relatively low on tracts of less than five thousand acres—there are four and a half million of them and together they constitute more than half the nation's commercial forest land; but despite considerable effort to improve management and raise output, little progress has yet been made. In Sweden, where small-forest programs have been emphasized for a number of years, productivity of large and small tracts is more nearly the same than in the United States. The experience there is clearly worth investigating.

Until less than twenty-five years ago there was every reason to employ all possible means of coaxing the American public to make more use of the National Parks; today park values in many places are being threatened by overpopularity. The outlook is gravest in the East, where the population is densest and there is no uncommitted federal land for new or larger parks. An especially acute problem is that of public access to beaches, which are nearly all in private hands. The experience of the English National Parks may offer useful clues; there, from the start of the parks program, it has been necessary to reconcile public enjoyment of park land with private ownership and use.

In the United States the preservation of natural areas has become a lively issue, largely because of the conflict between wilderness values of public land and pressures for economic and mass recreational use. In Great Britain this controversy seems to have been kept more within bounds and the use of natural areas for scientific observation and experiment is much more accepted and advanced than in this country.

As population grows, especially in urban centers, industrial and municipal pollution of water supplies increases even faster, creating new social, economic, and political difficulties. At least in the humid East, pollution has become the leading water problem. How much of the responsibility is federal and how much local? How should costs be divided among polluters and users who benefit by pollution abatement? In the heavily industrialized and thickly populated Ruhr district of West Germany, stream pollution is being held in check by an ingenious and complex cluster of local organizations that has been developing for more than fifty years.

Integration of federal land and water programs in the local areas where the actual work is done has vexed U.S. administrators, legislators, and landowners for a long time. Overlappings and sometimes conflicts in local efforts have resulted. Can these be reduced without loss of vigor and skill in the programs of national agencies? In recent years the Province of Ontario has developed a system of multiple-purpose land and water districts that may suggest at least some of the answers.

Economic development of entire regions has received much attention of late. Several approaches have been advocated and there has been a little small-scale experimentation, but not since establishment of TVA has there been a major innovation. A program now under way in the Lower Rhône-Languedoc area of Southern France is using a mixed public and private corporation in a long-range effort to build up an irrigation agriculture and improve rural living as an integral part of a broader economic development program.

In every case the emphasis is on the social values involved in defining resource problems, establishing general objectives, and organizing programs to attain them. Except for purposes of background, little is said of specific scientific or technological methods or detailed techniques of administration. Skills of that kind have a way of showing up as needed all over the world, often without benefit of special programs, at least in developed countries with strong traditions of free enterprise. The resources problems that cause the most trouble in those areas almost always have strong social and institutional elements requiring public agreement on goals and public choices among alternative lines of action. As Irving Fox has said of present-day water development issues, the real difficulties are not those of knowing how to build a dam or waste-treatment plant, or of obtaining suitable materials and manpower, or even of finding enough money. Instead, the big questions are whether public funds would be better used for some entirely different purpose; whether the benefits from a dam are worth the agricultural, urban, industrial, or recreational values that would be lost through flooding; and whether the expected gains would go proportionately to the individuals who bore the cost.

For these reasons all six of the subjects treated here are examined against a wide background. An introductory essay by an American authority in the field outlines the essential nature of each problem and its significance in the United States. A second essay describes and analyzes the experience of another country in dealing with its own form of the basic problem or situation. The authors of these essays are outstanding scholars or administrators (sometimes a combination of both) who from long familiarity can write authoritatively about their subjects. Two of them, in fact, are in charge of the programs described. In five of these essays the author lives in the country he writes about. The one exception is largely a technicality; Professor Fair writes of pollution abatement in the Ruhr district not only from long, first-hand observation but also from professional collaboration with the dean of the area's water engineers. In the third essay of each group another authority—again an American—relates the foreign experience to the situation in this country, drawing general conclusions on its significance as well as seeking to identify any specific approaches or methods that might be adopted for use here.

All but one of the examples are drawn from Western Europe, whence most of this country's institutions and ways of looking at things have been derived. People there have been saddled with most of the shared resource problems for longer than we have and the solutions they have worked out over the years tend to be more exportable for the U.S. market than those of countries with less similar patterns of thought and action.

The generalization cannot, of course, be pushed too far. The relationship is not just the simple one of a young country learning from an older world which has lived through all of the current U.S. difficulties. Sometimes the situation is just the opposite. Mass ownership of automobiles is just beginning to make a dent on outdoor recreation and land preservation problems in Great Britain, and on the Continent as well. As H. C. Darby points out, the national park system in the United States was very much in the minds of those who planned the British national system thirty years later. Philippe Lamour, in describing the development program in the Lower Rhône-Languedoc region, refers to the example of TVA. Nor can the United States any longer think of itself as a thinly settled country with plenty of elbow room for trial and error in land and water management. The 1958 population of the 48 contiguous states averaged about 50 per square mile. (This and succeeding figures are in round numbers.) While the U.S. estimate is still on the sparse side as compared with France's 210 people per square mile and Great Britain's and West Germany's 550, it is already above Sweden's 45, and has climbed to about a third of the 145 average for all Europe. And if one considers only the northeastern seaboard from Washington to Boston (all the territory of the coastal states from Maryland to Massachusetts) he finds that this large region, with its own problem of open space and pollution, has reached an average population density of about 350 per square mile, an impressive figure even by Europe's standards.

Still, there is no doubt where the weight of experience lies. In what the English still call the New Forest, as E. M. Nicholson points out in his essay, "Crown management for conservation has been continuous for over nine centuries." And one can move down 600 years in time and come only to the period when early settlers on this side of the ocean were beginning to clear tiny patches of the great North American forest and to use the rivers as highways. Yet in Europe, as Gordon M. Fair notes, the condition of the Emscher River in the Ruhr district "gave rise to complaints as far back as the 1600's."

Many other pertinent examples of resource management could, of course, be found in other lands. The decision to confine the original Forum lectures and the resultant book to land and water administration was made purely on grounds of manageability. This ruled out a number of questions in the fields of energy, nonfuel minerals, and marine resources on which the experience of other countries undoubtedly would provide other useful comparisons. Even within the fields of land and water, many interesting possibilities are not covered here. Those from the ancient Asian civilizations or Communist countries (or both) were passed over because of tremendous differences in institutions and outlook. These contrasts are formidable import barriers when the commodity one seeks consists primarily of goals and values rather than specific techniques. A number of other foreign examples with good prospects of domestic application were considered. These included flood-plain zoning in Switzerland, flood control along the lower Po in Italy, intergovernmental control of water projects in the Snowy Mountains of Australia, and land-use regulations to maintain family-sized farms in Denmark. These were passed over reluctantly only because a balanced group of six subjects seemed enough to be presented adequately in one lecture series and one medium-sized book. The topics finally selected are presented only as questions of significance to the United States; there is no implication that they are necessarily the six *most* important or urgent among the many resources problems of our times.

Even with the choice of examples limited to countries of relatively similar social and economic backgrounds, and presented by informed and articulate authors familiar with conditions in the country involved, it quickly becomes clear that scarcely any experience in resource administration is fully transferable from one national climate of institutions and thought patterns to another. Issues of federal-state relationships, which often complicate resources policies and programs in this country, have no real parallel either in Canada where the province is the dominant level of government in resources matters, or in France where the department is dominant in nothing. Few Americans yet have the Englishman's ingrained sense of responsibility for keeping the landscape tidy. Nor in the absence of 300 years of combatting pollution in a heavily populated area like the Ruhr is it likely that municipalities and industries in this country will agree overnight on a formula for sharing abatement costs. These and many other barriers to straight adoption are made clear by the various authors, all of whom keep their feet on the ground even while their vision ranges. But adaptation is another matter. The essays that follow suggest strongly that in each of the six cases there are approaches and methods that we in the United States will do well to think about and perhaps to investigate more closely.

Henry Jarrett
June, 1961 Editor, Resources for the Future

I ENGLAND AND WALES:

NATIONAL PARKS ADMINISTRATION

CONRAD L. WIRTH

The Nature of the U. S. Problem

H. C. DARBY

• *National Parks in England and Wales*

MAURICE K. GODDARD

What the United States Can Learn

1. PEAK DISTRICT NATIONAL PARK. The bare rolling moorland of Kinder Scout, looking down Fair Brook from the northern escarpment, an elevation of about 2,000 feet. This is now access land over which the public may wander.

2. NORTH YORK MOORS NATIONAL PARK. Afforestation at Troutsdale.

3. LAKE DISTRICT NATIONAL PARK. Tarn Hows car park near Coniston, Westmorland.

An Esso Photograph

4. PEMBROKESHIRE NATIONAL PARK. Refinery and marine terminal of the Esso Petroleum Company, Ltd., on the north shore of Milford Haven. The plant is situated in a depression so as to conceal it as much as possible from the surrounding countryside. Design and layout of the plant are also subject to suitable safeguards.

5. SNOWDONIA NATIONAL PARK. A camping site along the Afon Mawddach to the north of Cader Idris. In the distance is Penygadair, rising to a height of 2,927 feet.

British Crown Copyright

6. **EXMOOR NATIONAL PARK.** **Village of Brendon. "The British National Parks . . . include villages and even towns around the margins of upland areas and their valleys. Any national view of a Park system must come to terms with the local life of each Park."**

THE NATURE OF THE U. S. PROBLEM

■ CONRAD L. WIRTH

IN ENGLAND AND WALES as well as in the United States certain public lands have been set aside for recreation and refreshment, for the enjoyment of natural beauty, for the remembrance of the respective nations' past, and for the development of knowledge in the field of natural science. While the philosophy underlying the parks of the two countries is quite similar, the American National Parks differ in many ways from their British counterparts.

The older and larger National Parks of America are perhaps unique in that they were set aside *before* the lands involved were inhabited by modern man and before man had an opportunity to make significant changes.

We in America take pleasure in the belief that our National

■

CONRAD L. WIRTH is director of the National Park Service, U.S. Department of the Interior. Previously he was associate director of the Park Service in 1951 and assistant director from 1931 until then. Earlier he was in private practice as a landscape architect in San Francisco, a partner in a landscape architecture and town planning firm in New Orleans, and a member of the professional staff of the National Capital Park and Planning Commission in Washington. He was awarded the Pugsley Gold Medal in 1946, the Department of the Interior Distinguished Service Award in 1956, the Horace Marden Albright Scenic Preservation Medal for 1958, and, in 1961, the Rockefeller Public Service Award. Mr. Wirth was born in Hartford, Connecticut, in 1899, and received his B.S. degree from the University of Massachusetts.

Park System contributed in some measure to the establishment of areas of similar purpose in England. But adaptation of the National Park idea to the circumstances and needs of England has provided an example that can in turn help us in the United States as the country becomes more thickly populated and its economy further developed. America has few opportunities—and these will not remain open to us long—to set aside new National Parks similar to the areas established earlier out of primitive wilderness types of land. To provide for the needs of the future we shall have to come more and more to the idea that some park and recreation purposes can —in fact, must—be served concurrently with other uses, provided that there are appropriate special procedures to safeguard all major interests. This concept will have special application to State Parks and National Forests and to designated recreation areas in the future.

To follow this course does not mean the disruption or dilution of America's primeval National Parks. The fact that America, in meeting the expanding outdoor recreation needs, must look more to the combination of recreation with other compatible uses of the same land is even greater reason to preserve inviolate the true, primitive wilderness areas in the National Park System. Other qualified areas that are still available should be added while we can.

Preservation and use—at the same time! The paradox of that mandate is the root of many of our park administration problems and of widespread misunderstandings. We are to preserve the parks for the future, to spare them from damaging demands for material benefits for which this country has plenty of other, and better suited, lands. Parks have been likened to living museums where man comes to see nature, to study nature, to enjoy nature, but not to interfere with nature's processes.

But the term "park" like the term "museum" presumes man's use. Parks are for the use of the people; the human factor is fundamental in defining their purpose, even though man is not intended under park principle to consume the resources of the area for material benefit. With parks, the use is nonconsumptive; the resource never diminished. The products are the intangible, invaluable, essential values of inspiration and enjoyment.

But even such use can be damaging. How to balance such use against preservation? Nature can be, and often is, a fragile thing. Humanity can trample it to death while in the process of loving and enjoying it. Balancing preservation against use is always difficult—always an exciting challenge to a park administrator. I think that we in America are doing a good job of it. We are being helped these days by our Mission 66 program, in which the Congress is financing a special effort to make the parks better able to provide for the rapidly increasing number of visitors—which by the year 1966, the Park Service's fiftieth anniversary, we predict will pass the 80 million mark. However, our National Park System is at present neither large enough nor representative enough to insure National Park opportunities for the 350 millions of people who may live in this country by the beginning of the next century. Fine as it is, the System today is not adequately representative of our natural and historical treasures. Part of Mission 66 is to identify the remaining, eligible areas before they are gone.

In looking to the future, we in the National Park Service, like all park administrators, I surmise, are pinning our hopes for the increasing success of our preservation-and-use program on mounting public respect for the parks, and a more understanding and appreciative attitude toward our heritage of natural beauty and historical significance. The more the visiting public regards itself as trustee rather than owner of the parks, the better protection the parks will receive.

Our responsibility is to provide for a balance of appropriate uses in the National Parks. This balance of uses makes park administration a very complex form of land management, and, I am convinced, one of the most democratic. We must provide for the scientist and the motorist, the auto-camper and the wilderness seeker, the powerboat enthusiast and the wildlife photographer, the vigorous and the contemplative, the expert and the novice, all without interfering with each other and all without harming the basic integrity of the parks themselves. In short, we encourage the American people to use their parks appropriately in ways which they will enjoy, yet, to quote the law which created the Service, "to leave them unimpaired for the enjoyment of future generations."

Despite the many kinds of park use we must provide for, the National Park System is only one component in the system of recreational facilities which the United States has and needs. The National Park Service administers those areas of the country which are of such matchless beauty, scientific importance, historical or recreational significance, that their protection is the concern of the entire nation. Units of the National Park System go by many names; national parks, which usually contain a complex of scenic, natural, and even historic features; national monuments, often created to preserve a single noteworthy natural or historic object; and historical parks, battlefields, memorials, and seashores which fall under the criteria of national significance.

But the National Park System can and does fulfill only a share of the country's recreation needs. Other and related opportunities for recreation are provided in National Forests, National Wildlife Refuges, public domain lands and at reservoirs—which, though they provide for important material resources, yet yield a generous dividend of recreation. However, many of these lands are subject to short-time notice of change in purpose—to meet material uses—and thus alter their original conditions. State park systems and the park and recreation facilities provided by counties and municipalities are what might be called the basic or "grassroots" recreation resources of the nation, providing for the daily needs of our increasingly urbanized population. The adequacy of each system vitally affects the others, because park and recreation facilities everywhere in the United States are interdependent and must bear their fair share in making our leisure hours healthful, meaningful, and enjoyable.

It is when we look well into the future that the British experience—how the National Parks of England and Wales are conceived and administered—becomes especially pertinent. Contemporary problems and programs there may well be giving us a preview of some of our own problems in the parks some years hence; for as the density of our population approaches that of England the park problems of the two countries may develop more similarities than they have today.

NATIONAL PARKS IN ENGLAND AND WALES

■ H. C. DARBY

AS EARLY AS 1810 WILLIAM WORDSWORTH said that visitors to the Lake District, in northwest England, thought of it as "a sort of national property, in which every man has a right and interest who has an eye to perceive and a heart to enjoy." Since then awareness of the need to protect unspoiled countryside has grown with the progress of industrialization, and has become more and more vocal. The Census of 1851 showed that already 35 per cent of the population of England and Wales lived in towns of over 20,000 inhabitants. Many societies, local and national, were formed to protect rural scenery and to maintain or secure public access to uncultivated open spaces. In 1926, the Council for the Preservation of Rural England (the C.P.R.E.) was founded to coordinate these various efforts, and it has continued to act as the unofficial conscience of a nation confronted with the increasing disfigurement of its countryside.

■

H. C. DARBY is professor of geography at the University College, London, a post he has held since 1949. Earlier in his career as a geographer he was a Fellow of King's College, Cambridge, (1932-45) and a professor at the University of Liverpool (1945-49). He is a member of the Royal Commission on Historical Monuments (England) and of the National Parks Commission. No stranger to the United States, Mr. Darby has a long-standing familiarity with American parks and park people and problems. He was born in Wales in 1909.

In 1929, the C.P.R.E. submitted a memorandum to the Prime Minister which urged the Government to undertake preliminary inquiries with the idea of establishing National Parks. The result was an official Committee (the Addison Committee) which reported favorably in 1931. This report made reference to the Act of Congress which in 1916 had established a National Park Service in the United States; but it also pointed to the physical and economic circumstances of Britain that made impossible the adoption of methods similar to those of the United States and of the Dominions. Moreover, it stressed the difficulties that would confront any National Park authority that might be created. Not the least of these difficulties would be financial and also, so it said, the attitude of "those who think that any expenditure on the preservation of natural beauties of the country is unjustifiable." The economic circumstances of the ensuing 1930's certainly did not provide a favorable political climate for the flourishing of ideas about amenity; but some people, and not least the C.P.R.E., continued to voice the need for the establishment of National Parks.

The war of 1939-45 witnessed an increased self-consciousness about what was being defended. A Ministry of Town and Country Planning was established in 1943, and the scope of planning control was greatly extended. The new Minister asked Mr. John Dower to examine afresh the possibility of establishing National Parks. The Dower Report of 1945 is the most important document in the history of the National Park movement in Britain. It is lucid; it is comprehensive; and, in effect, it provided a blueprint for a National Park system. A committee was then set up under the chairmanship of Sir Arthur Hobhouse to examine the practical application of John Dower's proposals, and to consider in some detail which areas should be selected as Parks. The report of the committee appeared in 1947; and two years later, in December 1949, the National Parks and Access to the Countryside Act was passed, and the National Parks Commission inaugurated.

Before we consider the work of the Commission thus brought into being, two general points must be noted. In the first place, there is the small size of England and Wales, and the limited amount of unoccupied and relatively wild land. Space for the recreation of some 46 million people has to compete with more strictly economic needs such as mineral-working, afforestation and the provision of water-gathering grounds, to say nothing of agriculture itself.

What unoccupied and uncultivated areas there are lie mostly over 1,000 feet above sea level and, taken together, they amount to about 7 per cent of the total area of the country. Even this high land with practically no settlements is in private or semiprivate ownership. The word "semiprivate" is here used to describe "common land," that is, land subject to common rights (usually to graze animals) exercised by neighboring groups of people. Such common land has been described as "the last reserve of uncommitted land in England and Wales," and it is widely used by walkers and campers, although they have no legal right of access. The law relating to these "commons" is extremely obscure and complicated, and they have recently been the subject of an official inquiry. Whatever be the exact status of common land, it is certainly not that of public domain land from which parks could be created and, in effect, owned by a national authority.

The result is that the British National Parks, as they were proposed and as they have come into being, include villages and even towns around the margins of upland areas and in their valleys. It follows that any national view of a Park system must come to terms with the local life of each Park. Both the Dower and the Hobhouse reports fully recognized that the well-being of those who live and work in the Parks must always be a prime consideration. Not only farming but maybe even industry must flourish in a Park. As recently as May, 1959, the Minister responsible for planning could say that National

Parks would fail in their purpose "unless we can carry with us those who have their home and earn their daily bread within the National Parks and unless we can so administer the Parks that we win their support for the whole idea."

The second general consideration is the existence not only of a highly developed system of local government but of very comprehensive planning legislation. Various Town and Country Planning Acts (and in particular those of 1925 and 1932) had already provided, in theory at any rate, a measure of protection against the disfigurement of the countryside. When, in November, 1937, the Government was asked about the possibility of setting up a National Parks Commission, the reply stated that "Local Authorities have powers for the protection of areas of natural interest and beauty." Ten years later, the Town and Country Planning Act (1947) consolidated all existing legislation, and greatly extended and strengthened the powers of Local Authorities. It must be noted, however, that certain uses of land were excluded from planning control, and among these were agriculture and forestry. Thus a farmer is free to change from, say, dairy farming to mixed farming without control under this Act. Proposals for afforestation likewise do not require planning permission.

When the National Parks Act was passed in 1949 it in no way replaced the existing machinery of planning. The Commission it established is in no sense an executive body, vested with tracts of countryside which it administers. It is in essence an advisory body, charged with two functions:

> Preservation and enhancement of natural beauty in England and Wales.
>
> Encouragement of facilities for open-air enjoyment and for access to "open country" in National Parks. Open country is defined as consisting "wholly or predominantly of mountain, moor, heath, down, cliff or foreshore."

The role of the Commission is to see that these interests of amenity and access are not lost amid the clamor of many competing claims. It is but small exaggeration to say that to the unofficial conscience of the C.P.R.E. has been added the official conscience of the Commission. But in many ways its functions are in conflict; a thronging crowd can only too easily injure the very countryside which it admires and to which it is given access. The members of the Commission are appointed by the Minister responsible for planning (i.e., the Minister of Housing and Local Government), but the Act does not prescribe a fixed number. At the present time, the Commission consists of a chairman, a deputy chairman and eleven other members. Its total staff, all centered in London, amounts to thirty. The story of their work is told in the annual reports which the Act requires the Commission to make to the Minister.

SINCE 1949, IN THE COURSE OF TEN YEARS, THE COMMISSION delimited ten national parks, which together cover 5,254 square miles, that is, about 9 per cent of the total area of England and Wales. (See map on next page.) More than a quarter of a million people live in the ten parks. The delimitation involved prolonged negotiation with Local Authorities over such matters as boundaries and administration. The care of each Park is vested in a local park committee. For those Parks entirely within one county this is a committee of the county council, which (together with the county borough council) is the main unit of authority exercising local planning control throughout England and Wales. For those Parks that extend into more than one county, joint boards (which are executive) or joint advisory committees have been created. The expenses of administration are borne by the respective county councils. The size of each park committee varies between 12 and 27 members. Two-thirds of the membership are drawn from the locally elected representatives within the county or counties concerned, and the remaining third consists of nominees of the Minister as advised by the Commission.

Section 9 of the 1949 Act declares that, in preparing or altering its development plan, "the local planning authority

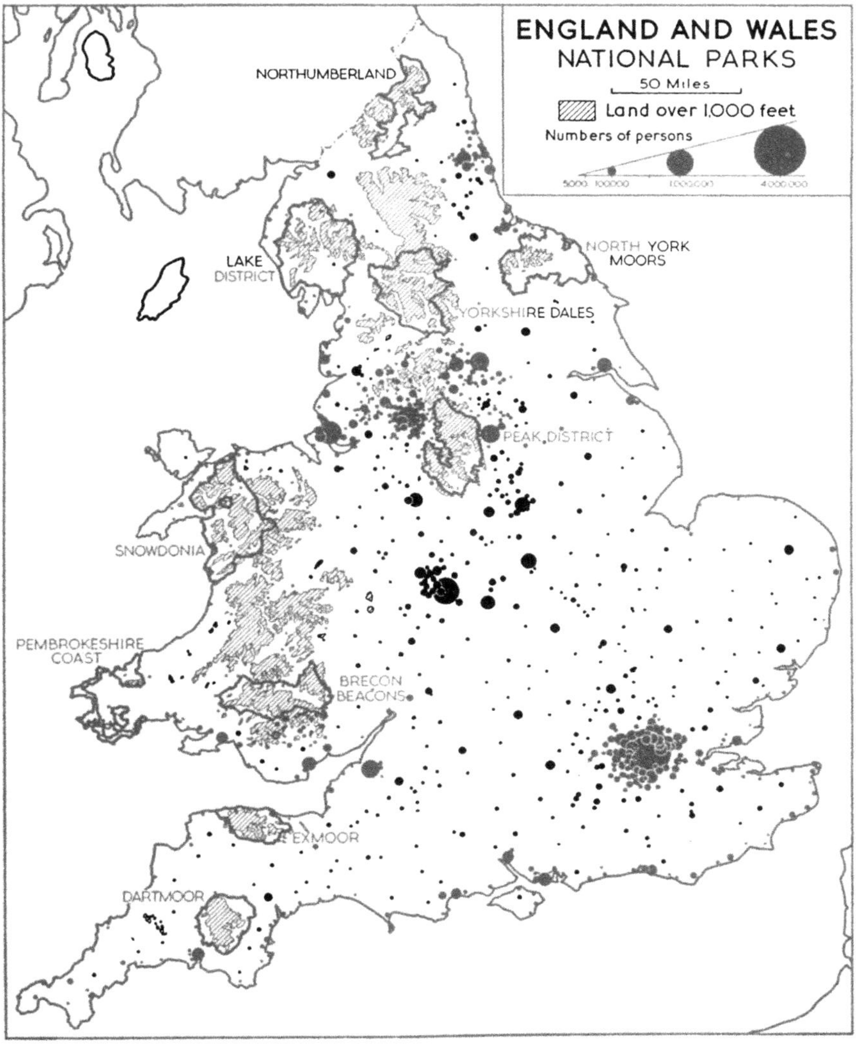

THE TEN NATIONAL PARKS OF ENGLAND AND WALES. The population of only boroughs and urban districts is shown.

shall consult with the Commission and take into consideration any observations made by the Commission"; the phrase "development plan" refers to the over-all plan prepared by each county under the 1947 Act. Close contact with the local park committees results in their agenda and minutes being provided for the Commission. In effect, the normal activities, services and planning decisions within each county apply as much to that portion within a National Park as to the rest of a county, with the proviso that the Commission has the duty of commenting upon major planning proposals within a Park—on the one hand to the local park committee, and on the other hand to the Minister concerned with planning.

From the point of view of a National Park, the considerations borne in mind by a planning committee, and by the Commission itself, fall broadly into two categories. On the one hand they are negative or preservative, that is, concerned with the protection of the existing landscape. Among the activities that might injure this landscape, the following categories will serve to give some idea of the issues involved:

> Industrial activities, such as quarrying and the establishment of power stations and oil refineries.
>
> Afforestation, whether by the Forestry Commission or by private companies and individuals. This generally involves the substitution of conifers for hardwood trees, and it also means the loss of "open country" for walking.
>
> The building of houses and the making of roads.
> The provision of aerial masts for broadcasting and for telecommunications, especially on high open ground.
>
> The construction of overhead lines for the transmission of electricity, and the degree to which it is economically feasible to place such lines underground where they pass through districts of very outstanding scenic quality.

The concern of the park authorities is not only with the preservation but also with the enjoyment and enhancement of the existing scenery. Section 10 of the Act requires each local park authority to notify the Commission each year of its program toward this end. The following five groups are examples of this kind of positive action:

> The provision of caravan sites, of camping sites, and of parking places for motor cars.
>
> The clearance of eyesores such as derelict buildings, derelict airfields, old mining sites, and derelict land generally.
>
> The establishment of information centers.
>
> The provision of access agreements to open country, under suitable arrangements that protect, for example, the shooting rights of owners or tenants.
>
> The establishment of warden services.

Section 97 of the Act provides that such positive actions are eligible for grants from the central exchequer of normally up to 75 per cent of the expenditure incurred. Many of the counties in which the Parks lie are, inevitably, areas of sparse population and so of low rateable value. The provision of even 25 per cent of the cost of capital projects, together with the annual expenses of administration, form a relatively heavy charge on local funds.

As the normal economic and social life of a park does not differ in principle from that of the surrounding district, it is inevitable that many proposals for development within a Park should come into conflict with the aim of maintaining the scenic beauty and rural solitude of this or that district. The local interests of those who live and work in a Park must be laid alongside the more national interests of those who visit the area for recreation and enjoyment. The inhabitants of a Park, faced maybe with local unemployment, and fearful of being denied some strengthening of their economy, may well raise the cry "Bread before beauty" or "Local interests before those of visitors." Nor is this duality of interest the only possible source of conflict. Powerful government departments, as well as private outside groups, may seek to promote developments that spell the very death of what the Commission seeks to promote. The interests of the Forestry Commission or of the Central Electricity Generating Board, for example, are not necessarily those of the National Parks Commission.

Conflicts of interest and difficulties of choice and judgment are best resolved by negotiation that results in voluntary agreement. Where this fails, the mechanism for breaking through the impasse is that of the Public Inquiry. This is not a method peculiar to National Parks. It is built into the planning legislation of the country, and many hundreds of inquiries are held each year over a wide variety of questions. In an Inquiry that affects a Park, one might reasonably expect that the interests of amenity will be strongly pressed.

A Public Inquiry can be instituted either when a private applicant for planning permission appeals against the decision of a local planning authority, or, when the Minister for planning (i.e., of Housing and Local Government) decides for various reasons to "call in" an application for planning permission and deal with it himself. When an official project is involved, an inquiry can also be held if the appropriate Minister decides that this is in the public interest; thus, electricity proposals are considered by the Minister of Power, road proposals by the Minister of Transport, military proposals by the Service Ministers. On the appointed day, in or near the locality involved, a representative of the Ministry concerned sits to hear the views of all the interested parties, who may number a score or more. In the following weeks he then considers the evidence placed before him, and, in the fullness of time, a decision is announced by the Minister, and this decision is final. Such a decision involves a balancing of interests. As far as National Parks are concerned, this may mean the balancing of amenity against, say, a local supply of electricity. The Chief Planning Officer of the Central Electricity Generating Board in an address in 1959 posed the question "What price for amenity?" And he went on to ask, "Is anyone brave enough to tackle the evaluation of amenity? How does one decide how high the standard shall be? How does one decide how much money—public money—to spend on safeguarding amenity?" In the same summer, Lord Strang, the chairman of the National Parks Commission, could speak of the "cruel dilemma" involved in this "conflict of opinion and conflict of interest."

The Peak District National Park. Some of these general points may emerge more clearly if we consider one example—that of the Peak District National Park which was the first Park to be designated, in 1951. It is set between two great areas of industrial population, that of the Manchester region in the west and that of Sheffield in the east; and there are also other areas of dense population nearby. It is thus conveniently placed for a day's visit or a week-end visit by walkers and cyclists, to say nothing of those who come by motorcar. The map on page 18 shows some of the features of the district.

The Park covers about 542 square miles; nearly 60 per cent of it lies over 1,000 feet above sea level and in places it rises up to or a little over 2,000 feet. Much of the higher and wilder country above 1,000 feet, especially that of the northern gritstone area, is moorland, often overlain with peat, and it provides water-gathering grounds for reservoirs that supply cities outside the park. Complicated negotiations with individual landowners have resulted in a total of about 37 square miles being made formally available for public access. As the moorlands have long been used for breeding grouse, the right of access is restricted, on this moor or that, for a limited number of days during the shooting season between August 12 and December 10. No shooting takes place on Sundays which is the most popular day for walkers on the moors. It must be noted that *de facto* access exists over a much wider area than that covered by access agreements. There is, incidentally, practically no "common land" in the Park, apart from village greens.

Villages and small towns are numerous in the lower-lying country below 1,000 feet and the map gives some idea of their distribution. Some 43,500 people live in the Park; some of these work beyond the boundary, for example in the quarries

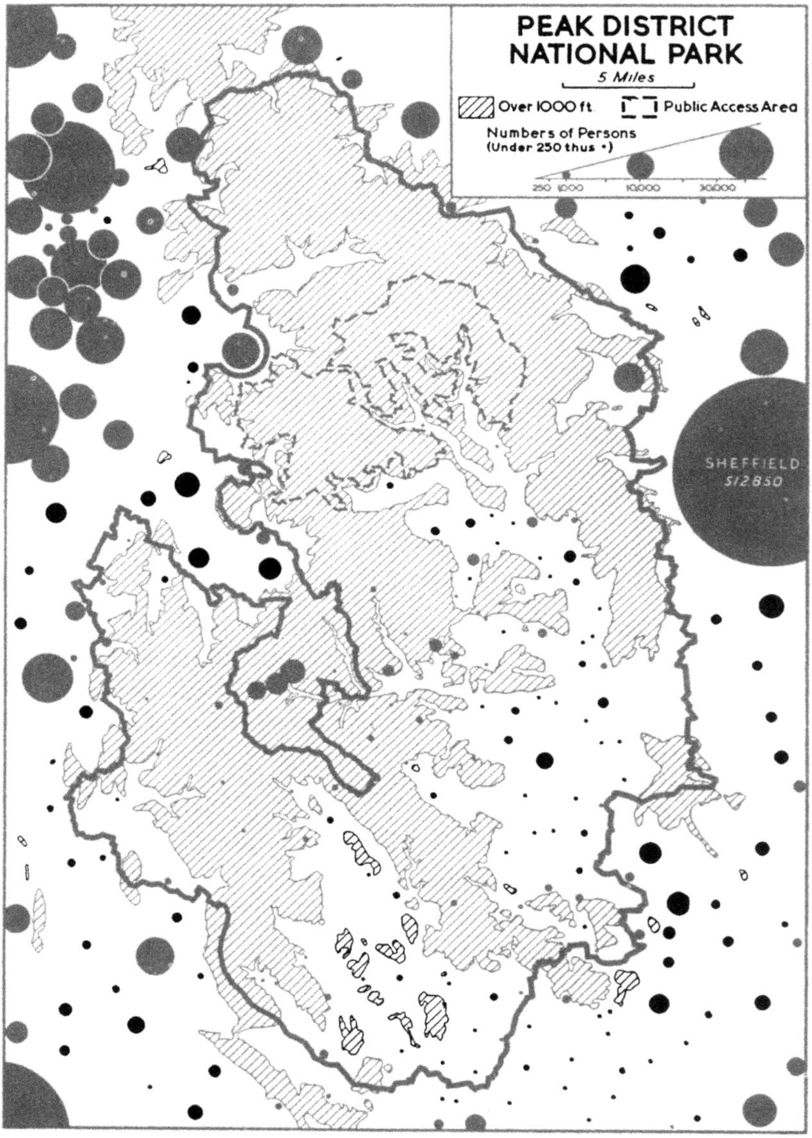

THE PEAK DISTRICT NATIONAL PARK. "Public Access Area" comprises (a) the area for which formal access agreements have been concluded, and (b) National Trust Land open to the public subject to the bylaws of the Trust. The population in each parish is shown by a single symbol, but the population of some parishes is widely dispersed throughout the parish territory.

that were deliberately excluded from the park, thus resulting in the curious re-entrant along its western border. Occupations of those who both live and work in the Park are estimated to represent the following proportions of the total number: agriculture, 32.4 per cent; mining and quarrying, 12.8; industries, 23.8; tourist services, 6.6; others, 24.4. The low figure for tourist services may be explained in part by the fact that two holiday centers (Buxton and Matlock) lie just outside the Park boundary.

Administratively, the Park stretches into four counties and into the county borough of Sheffield. Its Board consists of 27 members; 18 of these are appointed by the councils of these five authorities from among their elected members; the other 9 are appointed on the nomination of the Minister. Its total expenditure for the year ending March 31, 1960, including the cost of administration, amounted to nearly £40,000 ($112,000); but this figure is likely to be doubled in the near future. During 1959-60, the Board dealt with some 700 planning applications that included requests to build houses, to display advertisement signs, to make car parks, to erect telephone and electricity lines, and to mine for fluorspar. Permission was often given subject to conditions such as the use of appropriate building materials, the planting of trees to act as screens, or the placing underground of overhead cables in certain areas. Thirteen appeals against planning decisions were lodged with the Minister. There were also 79 cases of development being carried out either without permission or without a proper fulfillment of the conditions imposed. The general attitude of the Board in these matters is best indicated by a sentence from its annual report: "The Board's officers try to negotiate an agreed settlement in these cases if at all possible and authority for enforcement (i.e. compulsory) action is only sought as a last resort." During the year, enforcement notices were served by the Board in only six cases.

The total full-time technical and clerical staff of the Board is just over thirty. The Board also administers a warden service comprising two full-time paid wardens, four part-time paid wardens, and a large number of volunteers who are mostly members of various rambling and mountaineering clubs. Among its other activities the Board also maintains two information centers, provides traveling exhibitions, conducts anti-litter campaigns, organizes talks to school children, and issues various publications, including a handbook of accommodation and catering establishments.

This brief account does all too little justice to the day-to-day attention to detail that the Board's officers must give in this locality or that. Nor does it do justice to the sustained effort of protecting the landscape while promoting facilities for its enjoyment, and at the same time securing the interests of the 43,500 people for whom the Park is a home.

IN DISCUSSING THE WORK OF THE COMMISSION IT MUST BE said, in the first place, that the hopes of those who advocated the formation of National Parks have not been fully realized. As the chairman of the Commission said in 1959: "Certainly the Parks are not what the authors of the Hobhouse Report hoped they would be." The establishment of the Commission has not been sufficient to protect some of the most beautiful places in the land from injury. Assault has come from a variety of quarters. Four groups of intrusions have been selected for comment below—that of industry, that of aerial masts and pylons, that of afforestation, and that of caravans.

The assault from industry has been a powerful one. In 1957, the Commission opposed the establishment of an oil refinery, an oil terminal and an iron-ore stocking ground on the shores of Milford Haven in the Pembrokeshire Park; such developments it regarded as "inconsistent with the maintenance of the area as a National Park." But Milford Haven is, in itself, a great natural resource that was hitherto unexploited; it can easily accommodate ships of up to 100,000 tons deadweight. There was, moreover, some local unemployment, and the prospect of industrial development could not fail to make an appeal to many of the inhabitants of Pembrokeshire. "The balance of advantages in the national interest"—to use the Minister's phrase—led to the approval of each of these projects, "subject to the imposition of suitable safeguards" about the siting, the layout and the design of the installations. The employment of a landscape architect has certainly done something to mitigate the impact of the oil installations upon the landscape; up to the present, the plan for an iron-ore stocking ground has not been implemented.

Another major industrial proposal, in 1958, aimed at the establishment of an atomic power generating station at Trawsfynydd in the Snowdonia Park. At the ensuing public inquiry, the Commission urged the consideration of other sites, and said that "at the very least there must be the strongest presumption against the erection of large scale industrial installations in a National Park." But it also recognized "that there might be over-riding considerations and that it was for the Ministry to judge of these considerations." Later in the year, the Minister's decision gave consent for the construction of the power station "in view of the very strong technical and economic arguments for using this site." But the employment of a landscape architect was stipulated to reduce as far as possible the injury to "the natural beauty of the Park."

Another interference with the characteristic beauty of many areas is the construction of overhead transmission lines which march so unnaturally across a countryside, and in February 1960 a debate in the House of Lords drew attention to "the effect of electrical development upon the natural beauty of the British Isles." No one would wish not to have electricity, but there is surely a case for seeing that the new Industrial Revoluton should avoid the mistakes of the old. Much can be done to mitigate the effect of overhead lines by careful siting and by the placing of selected stretches of cable underground. Section 37 of the Electricity Act of 1957 expressly stated that the preservation of natural beauty should be taken into account; but the real problem is to what extent the Central Electricity Generating Board (which controls the nationalized industry) and the regional Area Boards, are prepared to lay aside annual sums for the placing underground of transmission lines that run through open country of especial beauty and interest. So far, to quote one speaker in the debate, "Section 37 has proved most disappointing."

With transmission lines, we may group aerial masts. The 1959 Report of the Commission expressed anxiety "at the large number of masts which were being erected in the countryside, not only for broadcasting sound and television programmes, but also to meet the telecommunications requirements of the armed forces, the police, other civic services and even industry." Much might be done by a policy of sharing masts, and some comprehensive policy needs to be devised with this in mind.

A third incursion into the traditional landscapes of England and Wales is afforestation. Since its establishment in 1919, the Forestry Commission has produced marked visual changes in many parts of Britain. The major problems that arise from these changes are concerned with the siting and design of plantations and with the species of trees to be grown. The early work of the Forestry Commission aroused much criticism because rows of conifers changed the character of well-known landscapes. Clearly, the interests of good forestry are not necessarily those of amenity, and the conflict between the utility of conifers and the attraction of hardwood trees is not easy to resolve. Arrangements have been evolved for consultation between the Forestry Commission, the Local Planning Authorities and the National Parks Commission; thus it is hoped to secure a balancing of interests. The Forestry Commission has, for example, agreed not to plant in the central dales and fells of the Lake District; this is a good example of the "modus vivendi" that has been established between the Forestry Commission and those interested in amenity.

More recently there has been a great increase in private afforestation as a result of concessions involved in the payment of both income tax and inheritance tax. Financial syndicates have been formed to invest capital in forestry, and the possibility of such planting in the Dartmoor and Exmoor Parks has provoked considerable discussion. As forestry, like agriculture, is not subject to planning consent, immense transformations could be effected in the scenery of many localities within the Parks without any check or control. Exactly what compromise will emerge between these interests and those of amenity remains to be seen.

A fourth incursion has come from the widespread growth of caravan sites in recent years. One popular form of holiday in Britain is to tow a caravan behind a car and seek out some convenient spot, more often than not near the sea, and so spend a relatively cheap family summer holiday. The increase in the standard of living in Britain is reflected in the increase of motorcars. In 1938 there were nearly 2,000,000 private motorcars licensed in England and Wales; by 1959 this figure had increased to nearly 5,000,000. The contrast between these two figures provides in itself an index of the increased pressure upon the remaining areas of open country. Not all who "caravan," tow their own caravans. Many people use those provided upon commercial caravan sites equipped with a piped water-supply and with sanitation facilities as well as with caravans themselves. It has been said that it is not economic to run a caravan site for fewer than fifty caravans. Planning approval for such sites has to be obtained unless "existing user right" can be claimed; but many sites carry more than their approved number of caravans, and there are many unauthorized sites. The recent "Caravan Sites and Control of Development Act (1960)" proposes a licensing system to regulate the establishment and operation of caravan sites; but it is too early to say what the effect of this Act will be. Caravan holidays constitute a valuable feature of summer recreation in Britain, but it would seem that the density of caravan sites in some localities is reaching a point when landscape of high value must be protected. The corollary of this is the provision of a sufficiently large number of authorized caravan sites near to but outside these critical localities.

The result of these four groups of intrusions, and of others, is that many examples of industrial and other inappropriate development can be seen in the Parks. But in "the balancing of interests," those of amenity do not always lose. The cumulative effect of decisions, small and large, on this point or that, in one Park or another, has been great. The achievement has been expressed by Mr. Harold Abrahams, the Secretary of the Commission, in a recent study of *Britain's National Parks* (1959) by twelve authors. He asked the question: How have the park planning authorities, backed by the Commission, succeeded?

> The National Parks themselves give the answer, but not the whole of it. The visitor cannot see the damage which has been prevented. He will not see in Snowdonia four major hydroelectric schemes proposed before the Park was designated, which would have adversely affected about a quarter of the Park. These were resolutely opposed by the Commission, and in the end they were not proceeded with. Nor will he see the blanketing of a wide landscape in the North York Moors by a large conifer plantation at Saltersgate on which the Commission made strong representations to the Forestry Commission some years ago. He will not see any overhead lines in Lower Borrowdale, one of the loveliest Lake District valleys, where the Electricity Board decided to put the proposed supply entirely underground after very strong protests against overhead construction made by the Commission to the then Minister of Fuel and Power. The wild open areas of the Chains in Exmoor will not now be clothed with conifers, thanks to the pleas put forward by the Park Planning Authorities and the Commission, supported by widespread public concern. Nor will the visitor see at Lee Moor, Dartmoor, china clay workings eating away a conspicuous moorland ridge on the south-west Park boundary, since the Minister of Housing, having heard evidence from the Commission and others, last year decided to contain the workings within limits more tolerable in the landscape. Nor the hundreds and hundreds of unsuitable smaller developments for which Park Planning Authorities have refused permission.

In all these instances "the balance of public advantage" has been in favor of protecting the scenery, and so of promoting the enjoyment of those who visit the Parks for recreation.

IN ASSESSING BOTH THE TASK AND THE ACHIEVEMENT OF THE Commission, one must recognize that National Parks cover nearly one-tenth of the total area of England and Wales. It would be both undesirable and impossible to divert such a large proportion of the country from making its contribution to the national economy. The interests of the Commission constitute one element in a democratic country with a dense population and with only a limited area of open country. Conflicts of interest and opinion are inevitable, and they must not only be resolved fairly but they must be seen to be resolved fairly. In the attempt to resolve these conflicts and to reconcile local with national interest, and private with public interest, five points might well be borne in mind, and these are set out below.

1. Whenever an industrial or other inappropriate development in a Park is proposed, all possible alternative sites should be explored before an assault is made upon the landscape. Once destroyed, the characteristic beauty of a fine stretch of country is irretrievable.

2. When the interests of amenity have to give way before other interests, as sometimes they must, it is important that the damage to the countryside should be minimized. The North Wales Electricity Act of 1952 provided for the appointment of a landscape consultant, and laid down that the local planning authority and the Commission should be consulted about details of siting and design. This amenity clause set a precedent, and many subsequent developments have been sanctioned subject to similar provisions. Appropriate siting and good design can do a great deal to mitigate the impact of industrial development and of caravan sites upon the countryside. The landscape architect who did so much to adorn England in the eighteenth century may yet have a wider, a more general, and a more vital, role to play in the twentieth.

3. The general financial situation of Britain has been difficult through much of the first eleven years of the Commission's life, and there have been frequent embargoes upon capital expenditure. The total amount of money spent on the work of the Commission has been absurdly small. The Tenth Report of the Commission (1959) could say:

> Over the last ten years, less than one penny per person *per year* has been spent in connection with National Parks. We do not believe that the public would be reluctant to spend, say, sixpence per person per year for the preservation and enhancement of our national heritage in the countryside. What could not be accomplished, if this extremely small contribution were forthcoming? For in a sense the preservation of the beauty of our countryside is a matter of pounds, shillings and pence.

Britain is obviously attempting to have its National Parks "on the cheap."

4. There may well be a case for examining each of the ten Parks, and for establishing "highest grade" areas from a scenic point of view. Caravan-free areas could be delimited; stretches of unspoiled coast could be defined; areas which might be afforested could be marked and also those which might not be. Some such scenic examination or survey might result in the definition within each Park of areas to be protected at all cost.

5. Finally, the impact of the Commission's work depends upon public opinion. And, in this respect, the Commission must lead as well as follow. Indications are not wanting that an increasing body of informed opinion is behind the work of the local park authorities. The Litter Act of 1958 and the Caravan Act of 1960, the insertion of amenity clauses in approvals given for industrial development—these are signs of an awareness of the threat to the countryside both within the Parks and without. But we must beware lest the Parks suffer from too little help given too late.

APPENDIXES

I—*Area and Population of the Ten National Parks*

Park and date of establishment		Area in square miles	Estimated population
The Peak District	(1951)	542	43,500
The Lake District	(1951)	866	44,000
Snowdonia	(1951)	845	30,000
Dartmoor	(1951)	365	28,500
Pembrokeshire Coast	(1952)	225	22,200
North York Moors	(1952)	553	23,000
Yorkshire Dales	(1954)	680	19,300
Exmoor	(1954)	265	14,400
Northumberland	(1956)	398	2,500
Brecon Beacons	(1957)	515	31,700
All National Parks		5,254	259,100
Totals for England and Wales		58,020	45,755,000

II—*Note on the Other Duties of the National Parks Commission*

In addition to the delimitation and oversight of National Parks, the Act of 1949 entrusted the National Parks Commission with other duties. Briefly, these fall into three groups:

(1) *The designation of Areas of Outstanding Natural Beauty (A.O.N.B.'s)*. These form the subject-matter of Sections 87 and 88 of the Act. They are areas which, for one reason or another, are not included in National Parks. They are usually without wide stretches of open country, and they consist largely of the ordinary agricultural land that makes up what someone has called that "sweet especial English scene." They are delimited in much the same way as the Parks themselves, that is, through negotiation with local authorities; and their designation has to be confirmed by the Minister responsible for planning. The work of delimitation began in 1955, and, so far, twelve areas have been designated, covering some 1,700 square miles. (See map on page 28.) Work towards the designation of other areas is in progress. Unlike the Parks, these areas have no special arrangements for

THE NATIONAL PARKS AND ASSOCIATED ESTABLISHMENTS. Areas of Outstanding Natural Beauty (A.O.N.B.'s) and Long Distance Routes are, like the National Parks themselves, administered by the National Parks Commission. National Forest Parks and, to a large extent, the New Forest, are administered by the Forestry Commission.

their administration. They fit into the general framework of planning control. As in the case of Parks, a local authority must consult with the Commission in preparing or altering its development plan in so far as an A.O.N.B. is affected. It must also consult with the Commission before entering into an access agreement or making an access order. The local authorities concerned with three A.O.N.B.'s have established special committees for them—Gower (The Glamorganshire County Council), Cannock Chase (The Staffordshire County Council) and the Malvern Hills (The County Councils of Gloucestershire, Herefordshire and Worcestershire). Approved expenditure in these areas is eligible for a grant of up to 75 per cent from the central exchequer. In effect, designation as an A.O.N.B. provides a measure of protection against unsightly development, but how effective this will prove to be depends upon the activity of the local planning authorities.

(2) *The establishment of Long-Distance Routes,* as required by Sections 51-58 of the Act. The intention, as stated in the Act, is that "the public should be enabled to make extensive journeys on foot or on horseback" along routes which do "not pass along roads mainly used by vehicles." Approved expenditure can be authorized for compensation, for the construction, maintenance and improvement of stretches of path, and for the provision of facilities for "accommodation, meals and refreshments."

The delimitation of these routes involves an enormous amount of most complicated negotiation about rights of way with local authorities and with private landowners. Proposals for establishing a route must be approved by the Minister. Work upon the Pennine Way started as long ago as 1951. It passes for some 250 miles through nine counties, and rights of way along all its length have not yet been secured. Work upon six other routes is proceeding—along the Welsh Border and along various stretches of coast. The total length of the seven routes amounts to about 1,000 miles, but much detailed work has yet to be done before they are in full operation.

(3) *General advice on natural beauty.* Section 85 of the Act places upon the Commission the "general duty" of advising "on questions relating to natural beauty" not only in the Parks and the A.O.N.B.'s, but anywhere in England and Wales. The Commission may thus comment on any development that is likely to injure the landscape, and it reports on questions referred to it by government departments, by local authorities, or by amenity societies. These are extremely wide terms of reference. In its eleventh year (1959-60), for example, more than 400 development

proposals were referred to the Commission; they concerned such matters as overhead electricity lines, mineral workings, caravan sites, and refuse dumps. Some of these matters were of only local significance. Others were of substantial national importance. Frequently, the Commission has helped to secure the insertion of conditions in planning approvals, thus minimizing the impact of development upon the landscape of many localities. As part of its general concern with the countryside, the Commission was required to prepare "a code of conduct for the guidance of persons visiting the countryside" (Section 86). The result was a booklet called *The Country Code* published at fourpence. Nearly 75,000 copies have sold, and the ten maxims of the Code have received wide publicity in schools, in broadcasts, in the publications of automobile associations, and elsewhere. The message of the Code has been reinforced by a series of posters designed to illustrate the importance of good behavior in the countryside.

III—*Note on Public Access to Open Country*

The first Access to Mountains Bill was introduced into Parliament in 1888 by James Bryce who later became British Ambassador to the United States. Bryce was a keen rambler and mountaineer, and his Bill aimed at preventing a landowner from excluding walkers from uncultivated mountain land. The Bill failed to secure support, and, during the next fifty years or so, other attempts to pass similar Bills were likewise unsuccessful. Much of the opposition to these attempts came from the owners of sporting rights, because grouse moors and deer forests had become valuable assets. It was, moreover, felt by some people that the urban public, unaccustomed to country ways, might do damage to stock, to trees, and to water supplies. On the other hand, feeling amongst many groups of walkers and ramblers ran high, and there were demonstrations against closed moors. In the meantime, access to many "commons" had been secured as the result of a long struggle supported by the Commons Preservation Society.

Partial success came in 1939 when the Access to Mountains Act was passed. This, however, was an unsatisfactory Act partly because it had been sponsored by a private Member of Parliament and so could not include a clause providing for monetary compensation, and partly because the procedure it laid down for securing access was complicated and expensive. In any case, the country was soon at war, and the Act never took effect. It was repealed and replaced by Part 5 of the National Parks and Access to the Countryside Act of 1949. This was concerned with "access to open country"; and "open country," as we have seen, was defined as consisting "wholly or predominantly of mountain, moor, heath, down, cliff or foreshore."

This Act of 1949 did not mean that access was automatically provided to all open country, but that every local planning authority was required to consider what steps should be taken to secure access to the open land within its area. Legal access could be secured in one of three ways: (1) by voluntary agreement (Section 64); (2) by the making of a compulsory access order to be confirmed or not by the Minister after the usual procedure of a public inquiry (Section 65); or (3) by compulsory purchase of the land (Section 76).

How has this worked? No compulsory access orders or acquisitions have been confirmed because policy has always favored negotiation rather than compulsion. The number of voluntary access agreements has also not been large except in the Peak District to which walkers come for a day or a week end from the two areas of dense population on either side. (See map on page 18.) As we have seen, the access agreements in this area are frequently subject to certain restrictions on a number of days during the shooting season. Part of the reason for the relatively small number of agreements may lie in the fact that the public already has *de facto* access to much open country, so that the impulse to clothe this access with legal form is not impelling. Thus the Lake District Board has decided that adequate access already exists, and that no further action is needed. On the other hand, the advantage of an access agreement for a landowner lies partly in the fact that the area may be made subject to bylaws that protect it (and its birds, fish and animals) from damage, and partly in the fact that the area may then be patrolled by wardens. This is a considerable advantage when one remembers that, under the common law of trespass, damage or nuisance has to be proved, and that proof in detail is not always easy to obtain. Fundamentally, the success of any access facilities, whether formal or not, depends upon goodwill between landowners, farmers, planning authorities, and visitors.

IV—*Note on National Forest Parks*

National Forest Parks are organized and administered by the Forestry Commission, and are quite separate from the ten National Parks; they are, moreover, not restricted to England and Wales but have also been established in Scotland. They are also quite different in character, for they are owned by the Forestry Commission and are not subject to planning control under the 1947 Act.

Realizing the difficulty of reconciling the claims of amenity and of economic utilization, the Forestry Commission (established in 1919) obtained powers under an Act of 1927 to make regulations governing the admission of the public to State Forests. That Commission also instituted in 1935 a Joint Informal Committee with the Council for the Preservation of Rural England (C.P.R.E.), for the discussion and settlement of conflicting claims. The idea of promoting access was enlarged into that of making Parks in some of the National Forests. Some paragraphs from the 1943 Report on *Post-War Forestry Policy* will show how the idea was worked out:

> When land is acquired for afforestation not all is plantable. Some of it (the arable land and improved pastures) is too good for trees, some (the tops of hills) is too rocky, too peaty, or too exposed for economic afforestation. The latter land is designated "unplantable" and but slight value is placed upon it when making acquisitions for afforestation purposes. Out of 1,144,000 acres of land acquired to September 30th, 1939, 370,000 acres were classified as "unplantable."
>
> In hill country the proportion of unplantable land increases and among the mountains it may be as high as 80 percent, although generally it is much lower. It is the purpose of the National Forest Parks to make this land available to the public for recreation and, at the same time, to provide cheap camping facilities at convenient places. Access to plantations will be allowed when there is no risk of damage from fire. The procedure in forming a National Forest Park is now well-established. A small Committee is appointed to examine and report to the Commissioners on the possibilities of a selected area and requirements in the way of camping sites and buildings. The terms of reference have been "to advise how the surplus and unplantable land may be put to use of a public character."
>
> Three such Committees have reported on the three Forest areas

in which Parks have already been established. When a National Forest Park has been established, a small Executive Committee, which includes representatives of local interests, supervises its general management. The local affairs of the Park are attended to by officials of the Commission who know the ground and local conditions intimately and one, the Camp Warden, is in charge of the camping grounds. There is also a large Consultative Committee, constituted of representatives of the Commission and of the bodies which are interested in the movement. This Committee meets from time to time as business requires.

Three National Forest Parks have been established in England and Wales, as shown in the map on page 28, and four in Scotland. They provide camping sites, with facilities such as shower rooms and cooking shelters, for an ever-increasing number of visitors.

The New Forest (like the Forest of Dean) is a Royal Forest that has survived from the Middle Ages. The law relating to it is extremely complicated, although much of it is now in the hands of the Forestry Commission, which permits camping. "With its extensive open heaths and recreational facilities," it was described in *Post-War Forest Policy* (1943), as "the nearest approach to a National Park in Great Britain." Paradoxically, it "has not been deemed necessary to declare it a National Forest Park although it was the prototype."

V—*Bibliographical Note*

1. An excellent account of the work of the National Parks Commission, and of each of the ten Parks, is given in Harold M. Abrahams (ed.), *Britain's National Parks* (London: Country Life Ltd., 1959) 151 pp. It contains contributions by twelve authors. Mr. Abrahams is the Secretary of the Commission. A brief convenient summary of the events leading to the passing of the 1949 Act is the pamphlet *National Parks and Access to the Countryside* (H.M.S.O., 1950).

2. An annual review of the work of the Commission over eleven years is to be found in the annual *Report of the National Parks Commission* (1950-60).

3. The following official reports give background information about the movement that led to the establishment of the Commission:

a. *Report of the National Park Committee,* Cmd. 3851 (H.M.S.O., 1931). The report of the Addison Committee.

b. *National Parks in England and Wales,* Cmd. 6628 (H.M.S.O., 1945). The report by John Dower.

c. *Report of the National Parks Committee* (*England and Wales*), Cmd. 7121 (H.M.S.O., 1947). The report of the Hobhouse Committee.

d. *Footpaths and Access to the Countryside,* Cmd. 7207 (H.M.S.O., 1947). This is a report of a Subcommittee appointed by the National Parks Committee in 1946.

e. *Gathering Grounds: Public Access to Gathering Grounds: Afforestation and Agriculture on Gathering Grounds* (H.M.S.O., 1948). This is a report of the Gathering Grounds Subcommittee of the Central Advisory Water Committee appointed by the Minister of Health.

4. An account of the origin of the National Forest Parks is given in chapter 5 of *Post-War Forestry Policy: Report by H. M. Forestry Commissioners,* Cmd. 6447 (H.M.S.O., 1945). The annual reports of the Forestry Commission give further details. For the New Forest, see *Report of the New Forest Committee 1947,* Cmd. 7245 (H.M.S.O., 1947).

5. For the problem of common land, see *Report of the Royal Commission on Common Land,* 1955-1958, Cmd. 462 (H.M.S.O., 1958).

WHAT THE UNITED STATES CAN LEARN

■ MAURICE K. GODDARD

PROFESSOR DARBY'S PAPER again demonstrates that each nation—with its own particular pattern of social and political mores—will meet its problems in its own way.

Neither the geography, the politics, nor the customs of the United States are conducive to the adoption of park policies identical to those currently employed in England and Wales. The two cultures differ and public policy has been, and will be, shaped accordingly.

This is true despite the fact that both countries, and most other highly developed nations of the world, confront the same basic problems spawned by urbanism and industrial growth.

■

MAURICE K. GODDARD is Secretary of Forests and Waters of the Commonwealth of Pennsylvania, responsible for the supervision of the state's forestry, park, and water resource programs. He formerly was director of the School of Forestry at The Pennsylvania State University. In 1959, he received the Patriotic Civilian Service Award of the U.S. Army Corps of Engineers for his work in co-ordinating Federal-State water programs. In 1960, he received the American Forestry Association's Conservation Award for his program to build a State Park within twenty-five miles of every Pennsylvanian. He is a former Council Member of the Society of American Foresters and a life member of the Soil Conservation Society of America. Mr. Goddard was born in Lowell, Massachusetts, in 1912. He holds a B.S. degree from the University of Maine, and an M.S. from the University of California at Berkeley.

In the more advanced nations we are now well on our way toward realizing more of the social benefits which can be derived from an industrial society. Having met, to an increasing extent, the basic economic problems of production and the distribution of income, we are on the threshold of being masters, rather than servants, of the profound economic revolution through which the world has been passing for the last two hundred years. We have the higher wages, the rising standards of living, the ease of transport, and the leisure which will enable us to exploit the benefits of industrial efficiency and more fully realize our individual capabilities.

Having come this far, we in the United States, as well as the people of Britain, now place a premium upon amenity. We seek relief from the routinized life of early industrialism. We seek the healthy exercise of power over the patterns of industrial and urban land use. We seek an environment in which a more enjoyable life is possible. We seek, in short, an aesthetic for twentieth century society.

In many respects this is one of the fundamental issues of our era. And since all nations share the liabilities as well as the potentialities of the new urban-industrial civilization, we can learn, one from the other.

The question before us therefore is this: Granting that both countries share a desire to provide the recreational and scenic amenities so important to the health and balance of an increasingly urban-oriented society, how applicable are the national park policies of Great Britain to our needs in the United States?

First, we must define what we mean by the term "park." Obviously the National Parks of England and Wales are not the same as Yellowstone or Yosemite.

As I understand it, national parks in Great Britain are not "owned" by the public. The parks are simply delimited areas within which a certain degree of control is exercised over economic development in an attempt to preserve the region's existing character. Within these areas are lands under several different ownerships, both private and communal. Villages or towns may even exist within their bounds or immediately on the edges. Consequently, the wish to preserve the landscape is subject constantly to the usual hazards of economics and of politics. Sometimes, as Professor Darby points out, national parks policy in the British Isles becomes a sort of rear-guard action against the steady encroachments of commerce and industry.

A second noteworthy characteristic of the British park system is the lack of intensive recreational development, with the exception of caravan (or trailer) camping sites. In England and Wales the National Park is a zoned preserve marked off primarily for its scenic amenities, but relatively undeveloped for the more active forms of recreation.

Traditionally, parks in the United States are owned by the people in fee title and, unless they are wilderness areas, must contain at least some recreational development.

The British philosophy is perhaps better fitted to a densely populated country with very little open space available. It may, in fact, be one of the few alternatives in a nation with a long, long history of settlement and exceptionally strong ties between the people and their land. It is certainly conditioned by the limited resource base available to 46 million people.

Does this mean, then, that we can learn little from their experience? I do not think so.

There is an intriguing possibility that some of the pragmatic and typically English techniques can apply to our own landscape, particularly in the older, more heavily settled regions of the United States where there are few extensive tracts of park land in public ownership and where present abusive patterns of land development threaten to snuff out the "livability" of the entire region.

Indeed, some of the basic principles of the British program already have been used in one way or another in various parts of the country.

Agricultural zoning, first implemented in California's Santa Clara County, south of San Francisco, has some similarity to parts of the British program in that its intention was to preserve the existing character of the land and thereby contribute to the general aesthetics of the region. It must be added, however, that the main intent of this zoning was to protect agriculture and not to provide the recreational amenities. An example of similar intention is New York State's policy for its Adirondack and Catskill Forest Preserves. In these vast tracts, the State of New York wants to preserve natural conditions to the point where even timber management is excluded. But here the difference between the two countries is one of method: the people of New York enjoy the great advantage of outright public ownership, and, consequently, encroachments are held to a bare minimum.

I know of no region in the United States where existing patterns of land use are being protected on a large scale through zoning and planning measures for scenic and recreational opportunities. The National and State Forests, the National and State Parks, and the large city parks are all publicly owned.

Wherever zoning has been employed in the United States, on even a limited scale, it suffers constantly from the steady and sometimes irresistible attrition of variances and revisions that in most regions have made it useless as a land control measure. There may be, of course, some exceptions I have not heard about, but I am sure the general statement is true. From the many sad experiences I have had myself I have been forced to conclude that there is no substitute for public ownership.

Even in the United States, land is too scarce and money too short, however, to justify any hope that the many large water-gathering, scenic, and recreational areas which should be protected from heavy urban and industrial incursions can be brought under absolute public control. Yet the preservation of such lands is essential in the ecology of a balanced urban environment. The watershed protection they provide, the relief they afford in a metropolitan landscape, the opportunities they present for recreation, and the role which they might possibly play in combating such complicated problems as air pollution all combine to indicate that it is high time for intelligent land-use controls in areas adjacent to the nation's huge metropolitan-industrial complexes. I am convinced that any workable program to achieve such control must rest in part on principles somewhat similar to those used in Britain.

I do not visualize such a scheme in the great open spaces of the West, but I do see it as one of several useful devices in the more crowded sections of the country where a degree of control must be exercised over land use in the interests of general economic and social welfare.

Take, for example, the Delaware River Basin, one of the most intensively developed in the entire world. The waters of this comparatively small basin—about 1 per cent of the area of the continental United States—supply the needs of 22 million people; it has one of the highest rates of industrial water withdrawal on earth. Part of the upper watershed of this important river is protected by the Catskill Forest Preserve. But as we come south to the Pocono Mountains of northeastern Pennsylvania we can see how some of Professor Darby's park philosophy might come into play.

The Pocono Mountains perform important watershed functions on the Upper Delaware. There lie the headwaters of the Lehigh and Lackawaxen Rivers, as well as many minor streams, which feed the Delaware system.

In addition, the Poconos are conveniently near the New York and Philadelphia metropolitan areas and the smaller urban centers of the anthracite coal fields and the Lehigh Valley. They serve as an important playground, therefore, for many millions of city people.

At the present time the Pocono region is devoted almost exclusively to forest and recreational use. About 30 per cent of the area is under public ownership in State Forests or State Parks. But these conditions are not immutable. A network of interstate highways will soon criss-cross the Poconos. Even today we are trying to protect the Bruce Lake Wilderness Area from transgression by a four-lane expressway.

While Bruce Lake is not a wilderness area in the strict

terminology of the National Park Service, it is wilderness to the people of the Commonwealth of Pennsylvania and it is so used. The lake is very small—several hundred acres—with 2,000 or 3,000 acres of land around it. There is no road—you have to walk in to it. You can fish in the lake but you have to carry your pole or your canoe. The Highway Department wants to put a four-lane expressway right smack through it in spite of the fact that the state owns it and already is using it for other purposes. If we didn't own it, I am sure it would not be worthwhile even to protest.

Spurred on by highway plans of this kind, industrial and commercial development will soon be pressing in on the forested highlands of the Poconos to the detriment of their present functions in the Delaware Basin.

The question is this: Which is more important in the long run, their water-gathering and recreational potential, or their availability as sites for new factories? There is little doubt that on scientific or engineering grounds the former kinds of use are the more valuable. How much longer can we adhere to a system of values which places a premium upon commercial growth to the exclusion of natural land functions or the amenities? Not much longer, in my opinion, for the tastes and the wants of twentieth century citizens can not be flouted without paying the price.

What can be done? How far can we go? Two years ago, at a State Forestry Association meeting at Bedford, Pa., Conrad Wirth, Director of the National Park Service, said that we might use perhaps the whole Appalachian Mountain region as a park. We face the dilemma, however, of outrageous expenditures if we try to protect such vast tracts of land through public ownership alone. Given present opinion, the American people would not stand for it. If on the other hand we entertain the notion that we can go in and protect an area such as the Poconos through a large-scale zoning proposal similar to that employed in Britain, we are politically naive.

I do not wish to over-stress the material side of the American people, but under our social system every citizen is eligible to share in the opportunities of economic growth, land holder or not. If he *is* a landowner and you wish to deny him the right to develop or sell his land for whatever purpose he wishes, you must compensate him. Without such an arrangement, large-scale zoning or any other proposal is doomed to failure.

It is entirely possible, however, that we shall be able to convince policymakers that some broad-scale control of land use is essential and practicable, provided we reimburse landowners for surrendering certain rights. What I have in mind here are the conservation easements or development rights about which there has been so much discussion in the last few years. At present, however, such policies are being considered for relatively small acreages. I would advocate a regional approach set up with an eye to all the needs of a river basin or a metropolitan complex. Such an approach can be a useful tool. Professor Darby shows that it works in England and Wales. It is not, however, the only way of preserving or resurrecting our landscape. Because the risk of caving in under heavy economic pressure is inherent in this sort of program, I still advocate public ownership wherever possible, however.

Legislation authorizing the acquisition of lands by urban areas in advance of actual current requirements for parks, reservoirs and plain open spaces is essential if we are to save the needed acreage from being engulfed by metropolitan growth. It is no coincidence that Britain's Town and Country Planning Acts were passed at about the same time as the United Kingdom's National Parks Commission was created. They are part and parcel of the same problem.

But we shall have a difficult time selling such a policy to the American people unless we can give it a price tag and unless we can point to the tangible benefits in income and goods which it would earn for the community, measured in a way that can be compared with costs. The value exists, as we all know, but a good job of computing its dimensions has never been done.

In the United States it is not enough to point out the aesthetic benefits to be gained from such a program. Legislatures and Congress demand something more measurable. That is why I take issue with Professor Darby's apparent distaste for multiple-use practices on park lands in his country. In Pennsylvania it is difficult enough just to protect a farm, let alone worry about how the man farms it. If I had to tell people in Pennsylvania that they shouldn't want to see rows of evergreen trees, I don't know what I would do. This one example points up the tremendous difference between England and Wales and the United States. In this country we would never fight over what kind of trees to plant. Some hundreds of years from now, as we develop our country, we probably will have the same problems that England and Wales do now, but not yet.

But for other reasons multiple use already is something of a problem in the United States. I am a bit weary of the running battle, in this country, between the so-called "wilderness" interests on the one hand and the "multiple-use" school on the other. There has been a notable refusal on the part of both sides to understand what the other is after. There is no inevitable conflict in purpose between the two and no reason why they cannot be reconciled.

To my mind, we will succeed in imposing sensible strictures on land use only when we can justify controls in the light of *all* the resource needs of the community. ("Community" in this sense could be as large as the whole United States.)

In Pennsylvania, we have tied our regional park program to multiple purpose reservoirs and multiple-use land management. For example, we are currently trying to acquire land for a new reservoir. Two hundred and forty families would have to be dispossessed; the cost would be in the neighborhood of $5 million. We propose to take about three times as much acreage as the reservoir would cover. The unflooded land would be in parks. We could not justify this kind of expenditure for recreation alone, but if we can tie it to some other resource use we can get tremendous support for it. We plan also to submit to the state legislature proposals permitting the purchase of conservation easements and the reservation of open lands through close co-operation between the state and local agencies. Similar arrangements are afoot in New York and California.

Whether there is a role in all this for the Federal Government as there is for the Central Government in Britain is a question I have not resolved in my own mind. Traditionally, planning and land use powers remain in the hands of the states, and there is little likelihood that this will be changed in the foreseeable future, although Senator Williams' bill to help communities preserve open space is a halting step in the direction of federal participation.

In the strict sense, therefore, there appears to be little applicability of the British system to the National Parks program in the United States but, in principle, there is some potential in the field of local and regional land use control. The major task is to convince the people and their representatives that a policy of sane land use control and planning adds to the commercial, agricultural, and industrial vitality of the community, as well as its desirability as a place where the twentieth century family will want to live.

II SWEDEN:

SMALL FOREST HOLDINGS

RICHARD E. McARDLE

The Nature of the U. S. Problem

THORSTEN STREYFFERT

• *Management of Small Forest Holdings in Sweden*

JAMES G. YOHO

What the United States Can Learn

1. HAULING BY HORSE. This is still the most common method of getting timber to the river bank or truck road. Hauling time is gradually being shortened by the extension of truck roads into the forest tracts, and the tractor is more and more substituting for the horse.

2. NATURAL REGENERATION OF PINE (Region III. See map on page 69.) The seed trees have been cut.

3. MIDDLE-AGED (ABOUT 60 YEARS OLD) STAND OF SPRUCE (Region IV). The pulpwood has been thinned.

4. ABOVE—Scarifying of soil before planting. Tractors are provided to forest owners and Management Subsidiaries by Forest Owners' Associations.
LEFT—But planting by hand is still universal because stumps, rocks, and debris will not permit use of machines.

5. ABOVE—Barking machine provided by a local Forest Owners' Association.

6. BELOW—Floating logs down a river.

THE NATURE OF THE U. S. PROBLEM

■ RICHARD E. McARDLE

BY WAY OF INTRODUCTION, and as background for Dean Thorsten Streyffert's paper on the management of small forest holdings in Sweden, I wish to comment briefly on the nature and significance of small forest ownerships here in the United States.

Let me first emphasize the economic and social importance of the whole forest resource in the United States. About one-third of our total area is forest land.

Forest cover contributes to the stabilization of soil and helps to regulate waterflow and improve water quality. Large areas of forest, especially in the West and South, provide forage for livestock. Practically all forest land and forest streams and lakes

■

RICHARD E. McARDLE has been chief of the United States Forest Service since 1952, following eight years as assistant chief. Previously he was, successively, silviculturist at the Pacific Northwest Forest Experiment Station, Portland, Oregon; dean of the School of Forestry, University of Idaho; director of the Rocky Mountain Forest and Range Experiment Station, Fort Collins, Colorado; and of the Appalachian Forest Experiment Station, Asheville, North Carolina. Mr. McArdle was a recipient of the distinguished service award, U. S. Department of Agriculture, in 1957; named in 1958 by the National Civil Service League as one of the top ten career men in government; and in early 1961 was one of five Federal Civil Service people to receive the distinguished Federal Civil Service Award. He was born in Lexington, Kentucky, February 25, 1899, and received his B.S. and Ph.D. degrees from the University of Michigan.

provide habitat for wildlife or fish that attract several million hunters and fishermen. A large part of our population uses the forest for these and other forms of outdoor recreation. About three-fourths of our total forest area is suitable and available for producing crops of timber. This commercial forest land, some 484 million acres, is about one-fourth of the area of our 48 contiguous States.

Annual value of the timber-products harvest is one forest benefit measurable in terms of money income. Timber cut for commercial use in 1954 was valued at about $1 billion on the stump. It is probably a good deal more than that now. Value added to that raw material by harvesting, processing, fabrication, transportation, and distribution of timber products amounted to another $18 billion. The total tangible and intangible value of all the resources and all the benefits of forest land far exceed these monetary estimates for timber alone.

Against this general background, let me now explain why small forest ownerships are so important from the standpoint of timber supply. Although they also are important from the viewpoint of the other uses of forest land, let us confine our attention here primarily to the small forest as a source of timber.

Forest holdings of less than 5,000 acres in size include 55 per cent of the commercial forest land in the United States. They contain about one-third of present sawtimber volume, and supply about 40 per cent of the present timber harvest.

About 50 billion board feet, or nearly half of total growth needed to meet future timber demands, must come from small holdings. This is as much as the present growth from all ownerships combined. It is more than double the annual growth that small holdings now produce. And within that total the growth of softwoods from these lands would have to be increased threefold if the projected demand is to be satisfied.

The magnitude of these goals is the main reason why improving the productive condition of this segment of our timber economy is so essential to meeting total future timber needs.

The estimated total needs that I have just cited are what would be required to maintain present per capita consumption of timber products, assuming that population of the United States by 2000 will be 332 million. Two of the current population projections are above that figure; only one is below it.

There are about 4½ million small forest holdings in the United States. Nearly 3½ million are on farms. Farmers, as a group, control twice as much commercial forest land as there is in the entire National Forest system—and close to three times as much as is owned by forest industry. The condition of recently cut forest land in farm ownership is far poorer than in any other class of ownership. Small forest ownerships average less than 60 acres in size. The average for farm forests is 46 acres. More than half of the 4.5 million small ownerships have less than 30 acres of forest land.

Time, of course, is the vital factor if we are going to get from small forest properties the production that will be needed in less than forty years. Fortunately, there is a foundation of public and private effort on which to build. This work has been gaining momentum to an encouraging extent. Nevertheless a huge accumulation of forestry work will have to be done on small forest properties in the next two decades if the growth goals for the year 2000 are to be met.

To get this job accomplished within the next two decades will require: (1) Reforestation at three and one-half times the current annual rate, (2) timber stand improvement at nine times the current rate, (3) a far more rapid shift from low-standard cutting to improved harvesting practices, and (4) reduction by at least half in losses still being sustained from fire, insects, disease, and other destructive agents.

How is a job of that size going to be done? One trouble is we still don't know very much about what motivates owners of small forest holdings. Foresters have tended to assume that money income is what all or most owners desire and that forestry must be sold on that basis. In general, this is a sound approach, but we are finding that in many instances it has little appeal. Some small forest properties are acquired by inheritance, some by mortgage foreclosures, some by gift, some are purchased for speculative value. Many are used simply as a place to live year-round or during the summer season. Some are held purely for recreational purposes.

A variety of public and private efforts already are under way to foster management of small forests. These include the various co-operative State-Federal programs for forest-fire control, forest-pest control, tree planting and distribution of planting stock, extension education, on-the-ground technical assistance to forest owners, and the small watershed program. The Agricultural Conservation Program provides direct Federal cost sharing to individual owners who plant trees or do timber stand improvement work on their land. Among a number of privately sponsored efforts are the American Tree Farm System, The Keep Green Program, Trees for Tomorrow, and Tree Farm Families. Education and technical services are provided by a corps of professionally trained foresters employed by many of the pulp companies and lumber companies. Similar services on a fee basis are being provided by the growing number of consulting foresters.

There is no way to measure accurately the effectiveness of any one of these many individual programs or of all of them put together. But we do know that in the aggregate they are not likely to get the job done in time to give us the timber we are going to need forty years hence. This is not stated in criticism, but as a simple fact borne out by appraisals of planting, protection, and productivity of recently cut lands in small private ownerships.

Before our grandchildren have reached our age, there will be twice as many people in this country as there are now. The pressures on land resources of all kinds will be far more intense than any we now have, even though some of us think they are intense already. Forest lands of all ownerships will be pressed into service—not for one specialized use—but for all uses to which they can legitimately be put.

Furthermore, future needs for timber will have to be met from a smaller forest acreage than we now have. The total now available will shrink considerably through losses to highways, power lines, pipe lines, reservoirs, pasture and cropland, and urban growth. We simply cannot afford to neglect any large acreage of forest land in small ownerships. Ways and means for getting small forest holdings under management must be found—and carried out.

The pressing need for larger production from small forest holdings in the United States, and the problems that we already are encountering in trying to achieve it, make Sweden's experience particularly pertinent to us in this country. Sweden is known all over the world as a leader in forestry and the small holdings there are extensive. Although economic, social, and political conditions in the two countries differ in many ways, Dean Streyffert's paper on management of small forest holdings in Sweden is not only of interest but of direct practical value in our effort to improve our own situation in the United States.

MANAGEMENT OF SMALL FOREST HOLDINGS IN SWEDEN

■ THORSTEN STREYFFERT

SEVERAL COUNTRIES, with quite varying natural and economic conditions for forestry, have experienced problems in the management of small forest holdings. This seems to indicate a certain correlation between the size of forest holdings and the level of management attained. The relationship, however, is not a simple one. The human factors involved often are of great importance. Also, in most countries many of the small forest holdings are closely integrated with farming, a situ-

■

THORSTEN STREYFFERT is dean and professor of forest economics at the Swedish Royal School of Forestry in Stockholm. Except for a two-year interval with the pulp and paper industry he has been on the teaching staff of the School of Forestry since 1930. Before that he was secretary of the government committee on forestry for two years, after fourteen years of service with the Swedish Forest Service. He has written widely in the field of forest economics in both Swedish and English. In 1945 he was consultant to the Food and Agriculture Organization. Dean Streyffert was born in 1892. He is a graduate of the Royal School of Forestry and the Stockholm University College of Commerce, and received his M.S. degree from Stockholm University.

ation which makes its impact on the management of farmers' woodlots. This tie between agriculture and the management of small forest holdings is quite important in Sweden.

From what is known about the management of small forest holdings in different countries, it would seem that this problem has been solved somewhat better in Sweden than in most other countries. A recent investigation based on Sweden's last national forest inventory indicated that there was no great difference in the management of the forests belonging to the three principal categories of owners: (1) the government, (2) the lumber and pulp companies, and (3) private owners, most of whom are farmers. In general, the forest lands in private ownership are small holdings while the lands of the other two categories are mostly in large holdings. The factors investigated were volume of growing stock and yearly growth per acre, distribution of growing stock between diameter classes, the quality of trees, the extent of reforestation, of thinnings and of noncommercial cleaning in the youngest stands, all considered according to differences in site class and other relevant factors relating to the different owner groups.

The results of this investigation were somewhat surprising to many in Sweden. It had been widely believed that even though the management of small forest holdings has been constantly improving since the turn of this century, the concurrent improvements on the big forest holdings had kept their level of management ahead. Although this investigation is not the final answer to the very complicated question of comparing forest management on big and small holdings, it certainly gives a reason for analyzing the evidence in an effort to explain why Sweden seems to differ from most other countries in this respect. It must be remembered, however, that the interplay of a given set of agencies will give different results in different environments, especially in consideration of the great importance of the human factor, of tradition, and of the general place of forestry in the national economy and in the economy of the forest owner.

A FIRST STEP IN ANALYZING THE PROBLEM OF SMALL FOREST holdings is to examine the reasons why the management of small forest holdings in most countries is poorer than that of the big holdings.

Precise assessment of the significance of the size of holdings to the efficiency of management is difficult, not only in Sweden but in most countries, because forestry on the smaller holdings is generally integrated with farming, whereas the larger holdings are in most cases either integrated with forest industries or else owned by the State or other public bodies. The influence of the size of the holding on the forest yield and on the economic outcome of forestry cannot, therefore, be distinguished from the influence of ownership. Discussion of the influence of the size of holdings must therefore include the influence of ownership with special reference to the integration of forestry with other economic activities.

However, it is necessary to try to distinguish between the influence of the size of the holding and of ownership. The size of holding has a bearing primarily on the *efficiency* of the management, whereas the ownership largely decides the *forest policy* of the owner or at least his general attitude in regard to the management of his forest property, inasmuch as this is a part of his total property, and thus should be dependent on the principles according to which he runs the enterprise as a whole if in fact he *does* apply a principle to the management of his forest property, which may not always be the case. But it is principle, when it is followed, that should largely decide an owner's policy in regard to investments in forestry.

To consider first the influence of the size of holding, it is a common opinion in Sweden that the principal drawbacks of the small holdings are the following:

Trained personnel and skilled labor cannot be employed except by the agency of public authorities or by co-operation between the forest owners.

Mechanization is difficult to apply economically on small

forest holdings, especially in regard to the more expensive kinds of machinery.

There is a scarcity of capital.

Small forest owners find it difficult to keep up with the results of research and technical developments in forestry.

The smallest wood lots and those of unsuitable shape—(long and narrow strips are the most common) —are a hindrance to the practice of rational silviculture.

These drawbacks are probably universal rather than a characteristic of Swedish conditions. They are liable to grow worse as technical developments and rising wages necessitate increased mechanization and strengthen the need for trained personnel and skilled labor. I think this is important to consider when looking forward.

In Sweden as elsewhere the integration of forestry and farming has its historical explanation in the early settlement of the land. Beyond that, there are several ways in which inclusion of forest land can add to a farm's income and general usefulness. In Sweden this has been so evident that the forest has been considered a more or less necessary support to the farmer, especially in the wooded regions of the country where farming is less profitable and the average size of the farm is small. Thus in the beginning of this century companies were prohibited by law from buying more forest land, and for several years acquisition of land by any one not going to stay on the land has been difficult on account of strong legislative restrictions.

On the other hand, it has been maintained that the integration with farming has led to negligence of the rational management of the appendant forest land.

To sum up, the following advantages and disadvantages of the integration between farming and forestry may be recorded:

Advantages

Labor and draft animals (or, to a growing extent in recent years, the tractor) on a farm can be utilized to the advantage of both farming and forestry because the slack season in farming coincides with the busy season in the forest.

Small or otherwise nonsalable trees that need to be thinned or cleaned out can be cut and used on the farm.

The small holding will allow of a very intensive management as the owner, if he is willing, can learn to know his forest in great detail. This introduces the important question of the personal interest of the owner, and of the factors influencing this.

Disadvantages

The farm has greater and more urgent claims upon the owner's resources of time, labor, and money than the forest because the cultivated land cannot be neglected without immediate consequences to income and living standard. Thus the forest is apt to receive a smaller part of the owner's total resources than it could claim if the owner tried to realize the greatest income from the combined farm in the long term.

The private owner, especially the small private owner, has a shorter range of view than the forest-owning industries and the state. This makes him reluctant to make investments in forestry, even if it can be shown to him that they are financially profitable.

The farmer has already one main occupation and cannot be expected to devote himself wholeheartedly to another at the same time. This should make him less suitable to manage a forest property, at least on his own. Here again the personal interest of the owner comes into the picture. This personal interest of the owner constitutes the somewhat irrational human factor in the management of small private holdings, which is a principal explanation of the greater diversity in the level of management attained compared to company and state forests managed by professional foresters in accordance with certain defined principles.

These advantages and disadvantages under Swedish conditions are probably universal, with the reservation that in other areas the seasons of the year may be more or less suitable than in Sweden for combined working of agriculture and forestry.

To sum up and appraise the impact of integration between farm and forest it would seem that investments in silvicultural measures should be less attractive to the farmers than to the other groups of forest owners dealt with here. One should therefore expect a poorer state of management in the farmers' forests in this important respect, provided counteracting agencies cannot be introduced. However, such agencies are employed in most countries where the farm forests are important.

In Sweden the commonly used means to bring about sufficient investments in farmers' forests and other small private forests are legislation, financial aid, and education. These measures are administered by public agencies and constitute together what may be called the public forest policy. This is done in the belief that the keeping up of investments in order to obtain a sustained yield from the forests is principally a public interest.

Raising the efficiency of the management of small forest holdings is primarily a question of financial profitability, and as such is left to the forest owners themselves. This has resulted in different forms of co-operation.

However, there is no clear boundary between activities to stimulate investment and to increase efficiency, inasmuch as the public agencies also contribute to better management, and the forest owners' co-operative organizations encourage their members to increase their investments in silvicultural measures. Some measures are clearly working both ways, as for instance the education of forest owners.

PUBLIC FOREST POLICY IN SWEDEN HAS EXERTED A DECISIVE influence in improving management on small forest holdings.

After a long period of regulated national economy in Sweden, as in the rest of Europe, the liberalistic epoch in the nineteenth century brought freedom also to the Swedish forest industries. Under the influence of the rapid industrialization in Europe the demand for Swedish forest products grew rapidly, with corresponding demands on the forests. This caused apprehension of a timber shortage and eventually resulted in the forest law of 1903 requiring reforestation after cutting. This law was passed only after much debate; at the time, much more than it would have been now, it was considered an intrusion into private ownership. However, the law eventually was generally accepted by the forest owners, and has filled its purpose.

By itself, however, the law would never have brought the desired result. A necessary complement has been education of owners in the economic importance of their forest lands and of the best ways of managing them. This important task was performed by the county forestry boards, the same authorities that supervised the forest law. The boards have been given a great deal of self-government, and they have enjoyed the confidence and co-operation of forest owners—chiefly the small owners, as the big forest owners have had their own trained personnel. It is the small forest owners who have turned to the personnel of the county forestry boards for advice and for supervision of forest work. Until recently, such services have been rendered at low cost but the principle has now been adopted that the forest owners shall be charged at actual cost.

The forest law of 1903 was amended in 1923 by the stipulation that vigorous forest must not be clearcut, and can be cut only by appropriate thinnings. In 1948 the economic aspect was introduced by a further amendment which defined vigorous forest as one that yields a return on the forest capital—growing timber and soil—at a rate to be decided upon by the National Board of Private Forestry. Another amendment at the same time provided that the reforestation obligation of a forest owner should be at a specified rate of financial return on the cost of reforestation as determined by the same Board.

This Board was established in 1941 as an intermediate organization between the county forestry boards and the Department of Agriculture. It has decided that the rate of interest in calculating the reforestation obligation of the owner should be 2.5 per cent and the corresponding rate in determining vigorous forest 3 per cent. These rates were below the current level of the lending market even when they were decided upon and are still farther below the market rate now.

The Swedish forest law has been of fundamental importance in molding public opinion and improving forest management, especially on small forest holdings; this type of direct regulation remains an essential part of public forest policy. However, it is now considered unlikely that further progress along this line can readily be achieved. In fact, the last amendment to the law has given rise to obvious difficulties of application. Furthermore, it is significant that a committee that recently was appointed to investigate ways and means of improving silviculture and management on private forest lands, and especially on small forest holdings, was not asked to consider any possible new amendments to the forest law. Furthermore, this committee did not find it advisable to increase the present very limited grants to silvicultural measures that are reserved for special cases. Instead, the committee strongly recommended a strengthening of the educational work of the county forestry boards, together with such associated activities as providing small owners with simple working plans for their forests. The aim would be to further develop interest and the knowledge of small forest owners about how the forest responds to different treatments. Thus each owner could find out the best ways of increasing the return from his forest land much as he now calculates how to increase the return from his agricultural land. Thus in the field of public forest policy, future progress will mainly depend on the strengthening of educational agencies.

AS PREVIOUSLY NOTED, THE SIZE DISADVANTAGE OF THE SMALL forest holdings is primarily a lack of efficiency. To offset this disadvantage different kinds of co-operation have been practiced in Sweden.

The co-operative movement among the small forest owners

is now about thirty years old. It has had the benefit of a well-developed tradition of co-operation between farmers in the field of agriculture and related industries. Without years of agricultural co-operation as a forerunner, the Forest Owners' Associations could scarcely have been able to break down the individualistic tendencies still inherent in the farmers at the beginning of this century.

There are at present two kinds of co-operative organizations among forest owners in Sweden. The eldest and still by far most common are the Forest Owners' Associations. There are 23 local associations, generally working within the boundaries of a county with a Central Board in Stockholm that takes care of their common interests, especially those of an economic and political nature.

The first Forest Owners' Associations were founded in the early 1930's. The next and more far-reaching stage of co-operation, which is in the nature of Management Subsidiaries to the Forest Owners' Associations, is of quite recent origin, as the first subsidiary of this kind was founded only in 1955. These have met with considerable success and represent a quite interesting and promising kind of co-operation.

Forest Owners' Associations. The primary objective of the Forest Owners' Associations is to realize the highest price the market will allow for the forest products, principally saw logs and pulpwood, delivered by their members to the forest industries and other buyers. The Forest Owners' Associations gather all the scattered items of roundwood cut by their members into bids of great commercial size and then negotiate about the price as parties of equal standing with the buyers. These negotiations are now generally made for rather big regions comprising a number of local Forest Owners' Associations and the corresponding forest industries. In this way it is possible for the small forest owner to sell small lots of wood—a few hundred cubic feet—that otherwise could not be put on the market. At the same time it saves the forest industries the cost of dealing with a great number of small sellers. This is such a great advantage to the buyers that they are willing to pay a commission to the Forest Owners' Association. The amount thus received is the principal source of income by which the Associations cover their expenses. Besides, their members have to pay an entrance fee to the Associations according to the acreage of their forest; these provide the working capital for the associations.

More recently the Forest Owners' Associations have extended their activities into several new fields. Most important is their well-organized program for helping members to overcome the disadvantage of small holdings in harvesting their timber. This is a field in which company and state forests have achieved rapid development towards mechanization and other forms of rationalization, prompted by increasing scarcity of labor and by rising wages. Even if the small forest owners to some extent do their logging themselves, they are increasingly dependent upon hired labor; the resultant rising costs tend to reduce the stumpage value from their fellings. The Associations have done much to provide the forest owners with suitable machinery such as motor saws, barking machines and heavy tractors for road building, etc., and if necessary also with skilled labor. Lately some of them have organized a service to take over the whole management of a forest property for owners who wish such assistance. Great efforts are further being made to inform and instruct forest owners about new tools and machinery; demonstrations of these at work are met with lively interest. Much can be lost by not cutting the trees into saw logs and pulpwood in the most economic way; consequently the instruction of the forest owners in this respect is given much attention. The proper storage of the logs in the woods and on the landings to avoid deterioration is another important aspect of logging given close attention.

The last important stage is the acquisition of processing industries by the Associations. The purchase and rebuilding of sawmills has been followed by the acquisition of pulp and paper mills in southern Sweden and lately by the building of a big cellulose mill with a capacity of 80,000 tons a year. Construction has been started on another mill of the same capacity, also in southern Sweden, and plans are on the way to extend these activities to northern Sweden.

In recent years the Forest Owners' Associations have also been active in stimulating their members in practicing good forestry. Suitable machinery for scarification of the soil to facilitate natural regeneration is rented to forest owners and trained planting crews are furnished. (This, by the way, is also done by the County Forestry Boards.) An intensive program of information is carried on around the theme that small forests belonging to farmers should not be less well managed than company forests. Prestige comes into the picture strongly here.

The Forest Owners' Associations have been a powerful agency in improving the management of small forest holdings. The small forest owners have been helped to increase income from their forests, not only by strengthening their bargaining position against the buyers, but also by making the harvesting of their timber crop increasingly effective and economic. The improvement of stumpage values thus realized should increase interest in silviculture to further increase the yield from the forests. The associations have also actively promoted good forestry with the small forest owners. This has been done in co-operation with the County Forestry Boards.

What are the possibilities and limitations of the Forest Owners' Associations for further improving the management of small forest holdings?

One important factor will be the degree of enrollment in those associations that can eventually be attained. At present the Associations have a total of 126,000 members. Among them they own 16 million acres, or 58 per cent of the total forest area belonging to private forest owners. Membership growth has now slowed down, and it does not seem likely that the associations will become very much bigger than they are now unless new activities can be initiated to arouse the interest of reluctant or adversely minded forest owners. There may be several reasons why forest owners do not join their associations. One is evidently the at least moral obligation for the members to sell their wood products through the associations. Some forest owners are against joining the associations on principle.

Even with a much greater enrollment in the Forest Owners' Associations, there still would be some limitations to their possibilities of overcoming the disadvantages of the small holdings, and the passive or even negative attitude of some forest owners to invest in silvicultural measures. The principal hindrance remaining is the scattered nature of the small ownerships which, furthermore, are generally split up into two or more separate wood lots. This prevents the concentration of forest work to certain areas each year, which is an important advantage to the big forest holdings in logging, transportation, or planting, and other silvicultural work. It also makes it difficult to employ at full advantage more expensive machinery as well as skilled labor and trained personnel. The solution to this remaining hindrance to good management of small forest holdings, so far as a solution can be attained by maintaining present ownership, has been sought by the establishment of Management Subsidiaries of the Associations.

Management Subsidiaries. Prices of forest products in recent years have remained at the same level or even declined, while wages of hired labor and other costs in forestry have been rising. Gradually the economic situation has become less favorable for the small forest owners who depend upon hired labor but do not have as much opportunity as the big forest owners to reduce cost by rationalization.

Even hired labor has been increasingly difficult to obtain for temporary work, such as logging, on small holdings; skilled workers require employment the year around, which they can receive on contract in company and state forests. Thus only unskilled labor can as a rule be had in the small forest holdings and at higher rates than for permanent employment. This means less efficient work at higher pay with the prospects that things will gradually grow worse.

In this situation a closer co-operation between the small forest owners than could be provided by the Forest Owners' Associations in their present form seemed to be the natural solution insofar as this could make conditions for rationalized management the same on the small holdings as on the big forest holdings owned by the state and the companies. As a result of such deliberations, the first Management Subsidiary was created in 1955 in Jaemtland County in northern Sweden.

The leading characteristics of Management Subsidiaries are the following: The Subsidiary in itself is no legal party. Its members sign a contract with the Forest Owners' Association of the region to be members of the Association and sell all their forest products through the Association. However, this may not always be compulsory. The Association in its turn places at the disposal of the members of the Subsidiary a specially employed trained forester, called forest inspector, who is usually a graduate from a rangers' school. This man, as a leader and advisor of the Subsidiary, should have thorough experience of forest management and an open mind for rationalization and co-operation. Besides the forest inspector, an assistant to him may be employed to supervise and direct the forest work. The Association employs forest workers on a permanent basis and provides housing for them. It also provides additional forest labor in the logging season when this is necessary. They are paid by the Forest Owners' Association and are considered as its employees. The members pay for this personnel only to the extent that they call for their services, otherwise not. In some Subsidiaries the members pay a yearly fee according to the acreage of their holdings besides a somewhat reduced fee for employing the services of the personnel. Any member of a Subsidiary who wants to do his forest work himself or to work for neighbors is, however, free to do so.

The connecting link between the Management Subsidiary and the Forest Owners' Association is a body of five to seven members, one of them being appointed by the Association and the rest being chosen by the members of the Subsidiary. Besides serving as a contact between the Association and the

Subsidiary, this body administers the special fund belonging to the Subsidiary which is intended to serve common purposes that cannot be charged to the individual members.

The Management Subsidiaries are organized in a very flexible way for serving the diverse needs of their members. The minimum contribution of each member is the fee going into the common fund, in the case referred to amounting to 2 per cent of the sales value of forest products sold, or 4 per cent of the stumpage realized if the sale has been done on the stump.

It is desirable that even such members who do not call for the services of the personnel of the Subsidiaries take part in the planning of the forest work on the co-operative level. Such planning, for instance, can take into consideration how logging by different owners should be located each year in order to minimize costs of transportation and of workers' rest houses, to mention only a couple of the important ways of eliminating the disadvantage of small and scattered holdings and helping the Subsidiary function as one big working unit. Planning on the co-operative level is co-ordinated with planning for the individual forest holdings, which is made more efficient by the aid of simple working plans recommending certain fellings and corresponding silvicultural measures.

Members who want to have their timber harvesting wholly done by the Subsidiary can arrange for this at a reasonable fee (about one cent per cubic foot), corresponding to the actual cost of the services to the Association. The cost incurred for labor will be charged to the member's account with the Association. He can thus receive the net income from his forest without any activity of his own. The necessary silvicultural work, such as planting after fellings, can in the same way be executed by the Association and charged to the account of the members concerned. In principle, a member never pays for anything in cash but all costs are charged to his account and in this connection reported in detail, thus enabling him to discuss with the forest inspector ways and means of rationalization. This is important as a means of keeping up the initiative and interest of the forest owners, which, insofar as it can be fostered, is the principal asset of the small forest holdings. A forest owner can even hand over the entire management to the leader of the co-operative. This is especially convenient for those not living on their property.

Investments in planting and other measures of silviculture are greater in the Management Subsidiaries than in other private forests in the same region. In Jaemtland county such investments in 1959/60 amounted to about 40 cents per acre a year. They represented about 15 per cent of the stumpage value realized from fellings in the same year. The corresponding investment for the total forest area in the county as a whole is about 7 per cent of stumpage value. Thus the County Forestry Boards need not care for the silvicultural work in the Subsidiaries, as these have their own personnel.

To summarize, the Management Subsidiary represents an extension of the service rendered by the Forest Owners' Associations to its members. The members of the Subsidiary as a rule are also members of the Association and thus have at least the moral obligation to sell all their forest products through the Association. The flexible nature of the Subsidiary allows a great margin for meeting individual needs and wishes. At the same time it leaves the opportunity for members who so desire to manage their own holdings, but with full opportunity to discuss their problems with a professional man. This explains the great attraction this kind of co-operation has received. In the short time Management Subsidiaries have been in existence their growth has been rapid, especially for a period devoted primarily to trying out this new form of co-operation. At the present time there are fifty-two Management Subsidiaries with a total forest area of over half a million acres and with a total membership of 3,200 forest owners. Thus the average Subsidiary has sixty members with a total forest area of about 11,000 acres. Altogether about 400 permanent forest workers are employed.

The experience gained so far is encouraging. The fundamental problem of combining the advantage of the big forest holdings with the valuable self-activity of the individual owners seems to have found a solution in the Management Subsidiaries. They appear to offer a solid foundation for the management of small forest holdings in private ownership.

In the nature of things there will be limitations to this form of co-operation too. There will always be forest owners who for different reasons will not join a Management Subsidiary just as they will not join a Forest Owners' Association. However, many forest owners not joining the Subsidiary at the start joined it later, when they found out how useful it could be for them.

An important limitation to the potential spread of the Management Subsidiary as an idea for promoting management on small forest holdings is that a Subsidiary cannot be started just anywhere as an isolated unit and be expected to succeed, regardless of the environment. It is fundamental that it should fit well not only into the natural and economic conditions of an area, but also into its past social and political development, taking the shape of tradition and molding the minds and attitudes of man. One must take full account of environment as a necessary background for judging the possibilities and limitations of Management Subsidiaries in Sweden and possibly to some extent elsewhere.

THE APPENDIX THAT FOLLOWS BRINGS TOGETHER SOME OF THE basic facts and figures on forestry in Sweden, with special emphasis on small holdings and their relation to the rest of the economy. I hope that this information will give readers not already familiar with Sweden a better understanding of the problems and programs that I have discussed in this essay. I also hope that it will make it easier for persons in the United States and other countries to determine how, and to what extent, Swedish principles and methods might be useful in solving their own problems of improving the management of small forest holdings The relative importance of forestry to Swedish farmers, the proportion of productive land in forests, the fact that there are only three commercially important species of tree, and other physical or economic characteristics have a large bearing on how Sweden's experience might be adapted elsewhere.

APPENDIX

A Note on State Assistance For Private Forestry

County Forestry Boards. The County Forestry Boards are financed by the Government and by a special tax levied on all forest land. In 1960/61 Government appropriations amount to SKr 14.4 million (U.S. $2.7 million) not counting allocations for subsidies. The special tax in 1959/60 brought SKr 12.7 million (U.S. $2.2 million).

The Government appropriations (SKr 14.4 million) are intended to cover the cost of administration, supervision of the forest law and educational and training activities for forest owners and forest workers as well as a number of activities pertaining to public forest policy, including the planning of subsidized forestry work. Advice provided at the request of forest owners in connection with their own forest holdings and planning and supervision of forestry work on such lands are in principle paid for by the forest owners themselves. Until recently the small forest owners have not had to pay the whole cost for such services, as the special tax mentioned above has been used to help in financing such activities by the County Forestry Boards. However, this tax is now being abolished by steps, and accordingly the forest owners will eventually have to pay the whole cost of the services of the personnel of the Forestry Boards. This cost has recently been raised to SKr 80 (about $16) a day for trained personnel of a ranger's standing.

Subsidies. In principle no subsidies are given for the reforestation work needed for keeping up a sustained yield. On the other hand, subsidies amounting to 50 per cent of cost may be granted for the increase of the sustained yield, as by the afforestation of bare lands. In accordance with the same principle subsidies are also given for the draining of marshy lands (25 per cent of cost) and for the construction of truck roads in the forest (40 to 50 per cent). Re-

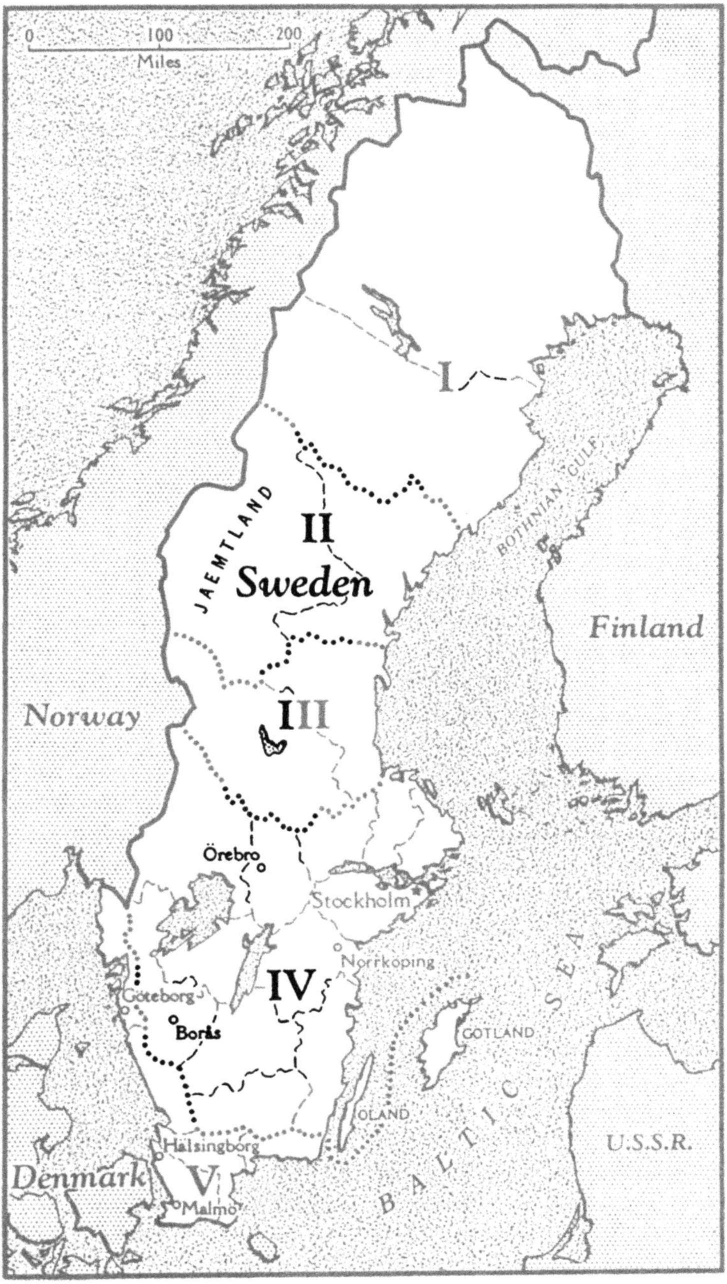

SWEDEN'S FIVE FOREST REGIONS

cently subsidies have been given for the afforestation of marginal agricultural lands. In special cases when conditions are unfavorable, subsidies may also be awarded to help financing obviously unprofitable measures of reforestation in northern Sweden required by law. These subsidies are intended chiefly for small forest owners.

The entire area of private forest land (not including company lands) is 28 million acres. The total cost of reforestation and other silvicultural measures (draining of marshy lands included but not road construction) on this area for the present amounts to about $5 million a year. The total cost of road construction was about $3.5 in 1960/61.

Economic Data for Hammerdal Forest Management Subsidiary in Jaemtland County

Forest area—24,468 acres.

Membership—98 forest owners.

Percentage of all private forest owners in the area—89%.

Percentage of total private forest area—88%.

Trained personnel employed by the Forest Owners' Association—one inspector (graduated from ranger's school) and one assistant.

Number of permanently employed forest workers—18.

Timber cut and sold to the Forest Owners' Association—1957/58, 537,894 cu. ft.; 1958/59, 299,758 cu. ft.; 1959/60, 391,595 cu. ft.

Total value delivered at river bank of timber sold in 1959/60—$98,620 = 25.2 cents per cu. ft. (without bark).

Total logging cost including overhead to river bank—$52,960 = 13.5 cents per cu. ft.

Stumpage realized—$45,660 = 11.7 cents per cu. ft. corresponding to 8.1 cents per cu. ft. of standing timber (with bark).

Cost of reproduction and other silvicultural measures—$15,090 (of which $1,760 performed by the forest owners themselves). The total cost amounted to 62 cents per acre and 33 per cent of the stumpage realized from sale.

Cost of truck road construction—$2,110.

Cost of personnel employed by the Forest Owners' Association (one inspector and his assistant)—$7,240. This cost is defrayed by a fee of 1.1 cents per cu. ft. of timber logged by the Management Subsidiary and by a fee of $12.40 per day for the services of the inspector or his assistant in marking trees for cutting, and advice and supervision of silvicultural work. The average income of forest labor by piece work in logging 1959/60 was about $9.00 per day in this region. Wages paid per hour are lower.

TABLE 1. Land area of Sweden by major classes, 1956

Classes of land	Northern Sweden (six northernmost counties)	Southern Sweden (except six southernmost counties)	Six southernmost counties	Total Sweden
Total land area (1,000 acres)	67,327	27,670	6,590	101,587
Per cent of total in:				
Productive forest land	56.2	59.0	37.8	55.8
Agricultural land	2.8	22.7	40.0	10.6
Unproductive land	41.0	18.3	22.2	33.6
Forest area per capita (acres)	25.9	4.2	1.5	7.7

TABLE 2. Distribution of farmland with over 12.5 acres of agricultural land in different regions of Sweden between agricultural and forest land (1951 Census)

Region and farmland by agricultural area	Number of farms	Farms without forest land	Farms with forest land of:			
			12.5 to 60 acres	60 to 250 acres	250 to 1,000 acres	1,000 acres and over
Southern Sweden (agricultural regions)						
12.5 to 25 acres	21,109	8,691	11,165	1,226	26	1
25 to 50 acres	29,266	11,513	13,995	3,635	121	2
50 acres and over	25,158	10,041	7,855	5,800	1,224	238
Total	75,533	30,245	33,015	10,661	1,371	241
Southern Sweden (wooded regions)						
12.5 to 25 acres	38,494	5,029	18,502	14,153	801	9
25 to 50 acres	19,831	3,509	7,222	8,149	933	18
50 acres and over	6,119	1,643	1,115	2,286	939	136
Total	64,444	10,181	26,839	24,588	2,673	163
Northern Sweden						
12.5 to 25 acres	25,399	1,084	7,072	13,223	3,792	228
25 to 50 acres	7,246	234	999	4,065	1,825	123
50 acres and over	650	26	22	243	310	49
Total	33,295	1,344	8,093	17,531	5,927	400

TABLE 3. The average farm in different regions of Sweden and its income from agriculture and forestry, 1958[1]

Region	Farm acreage			Net income[2]		
	Agricultural land	Forest land	Total	Total	Agriculture	Forestry
	acres	*acres*	*acres*	*dollars*	*per cent*	*per cent*
Southern Sweden (agricultural regions)	58	69	127	3,897	87	13
Southern Sweden (wooded regions)	29	95	124	1,881	60	40
Northern Sweden	22	183	205	1,601	46	54
Total Sweden	38	108	146			

[1] Includes only farms with more than 12.5 acres of agricultural land and additional forest land.

[2] Net income includes consumption of products from the farm. Interest on investment not deducted.

TABLE 4. Productive forest area in Sweden distributed by ownership (1951 Census)

Region	Percentage of regional total in:				Total
	State forests	Other public forests	Company forests	Other private forests	
	per cent	*per cent*	*per cent*	*per cent*	*1,000 acres*
Northern Sweden (six northernmost counties)	24.7	6.2	29.4	39.7	36,633
Southern Sweden (except six southernmost counties)	6.6	7.0	17.8	68.6	16,426
Six southernmost counties	5.5	5.5	2.9	86.1	2,505
Total Sweden	18.5	6.4	24.8	50.3	55,564

TABLE 5. Total volume of growing stock in Sweden (national forest inventory 1953-58)

Region	Total volume (with bark)	Volume per acre	Pine	Spruce	Broad-leaved	Less than 6″	6″-10″	Over 10″
						Diameter at breastheight		
	million cu. ft.	*cu. ft.*	*per cent*	*per cent*	*per cent*	*per cent*	*per cent*	*per cent*
I	15,278	895	47	36	17	34	41	25
II	13,396	1,260	32	54	14	34	42	24
III	12,662	1,490	47	43	10	27	43	30
IV	27,432	1,680	38	48	14	21	40	39
V	3,431	1,440	30	40	30	22	36	42
Total Sweden	72,199	1,310	40	45	15	27	41	32

TABLE 6. Maximum productive capacity of forest land in Sweden and yearly growth (national forest inventory 1953-58)

Region	Productive capacity		Yearly growth	
	Total	Per acre	Total	Per acre
	million cu. ft.	*cu. ft.*	*million cu. ft.*	*cu. ft.*
I	606	35.75	374	21.45
II	470	44.33	385	35.75
III	462	54.34	429	50.05
IV	1,246	75.79	1,067	64.35
V	202	82.94	160	65.78
Total Sweden	2,986	54.34	2,415	44.33

TABLE 7. Income and cost figures from six forestry areas in Jaemtland County served by Management Subsidiaries

Item	Unit	1957/58	1958/59	1959/60
Members' forest area	acres	106,300	130,500	146,800
Cut for sale	1,000 cu. ft.	2,710	1,890	2,432
Returns from sales (delivered at river bank or truck road)	dollars	732,000	451,000	620,000
Cost of logging, including overhead	dollars	363,000	269,000	295,000
Stumpage realized	dollars	369,000	182,000	325,000
Per cu. ft.	cents	13.6	9.6	13.3
Per acre	dollars	3.46	1.39	2.20
Cost of silvicultural measures (planting, drainage, etc.)	dollars	18,100	26,000	52,300
Per acre	cents	17	20	36
Per cent of stumpage	per cent	4.9	14.3	16.2

WHAT THE UNITED STATES CAN LEARN

■ JAMES G. YOHO

BECAUSE DEAN STREYFFERT HAS COVERED the ground so thoroughly, and because I am in basic agreement with the ideas he puts forth, I can come quickly to my main point: to examine the possibilities of applying in the United States the Swedish experience of dealing with the problem of the small forest ownership.

It is quite obvious from Dean Streyffert's presentation that forest regulation plays a very important role in Swedish forest policy; and, therefore, a discussion of other policy measures would not be complete without giving due consideration to its influence. However, inasmuch as forest regulation is a well worn subject to most of you, I shall omit it from this discussion

■

JAMES G. YOHO is professor of Forest Economics at Duke University. Previously, he was assistant professor of forestry at Stephen F. Austin State College (Texas) (1948-53) and assistant and associate professor of forestry at Iowa State University (1953-57). Intermittently since 1948 he served for short periods as a research forest economist with the Southeastern Forest Experiment Station and as a consultant to forest industries in the South. He is a member of the American Economic Association, American Farm Economic Association, and the Society of American Foresters. In the last named organization he is national vice chairman, Division of Economics and Policy. He is a member of the Governor of North Carolina's Committee on Area Development. Mr. Yoho was born in Brownsville, Pennsylvania, in 1920 and received his B.S. degree from the University of Georgia, his M.F. from the New York State College of Forestry, and his Ph.D. from Michigan State University.

except for a brief passing tribute to Sweden's way of handling the matter. In this regard, I doubt that one could do better than to offer the following quotation from Greeley's book on forest policy concerning forest regulation in Sweden which states that "This legal testing of silvicultural operations by an economic yardstick is probably the most advanced public regulation of private forestry yet undertaken in any country."[1]

Another significant omission from my discussion is the role of public subsidies to private forest owners. While this is, I am sure, an important aspect of Swedish forest policy Dean Streyffert does not touch on it in his paper. Accordingly, I shall do no more than make mention of it here.

At the outset of his paper, Dean Streyffert emphasizes that tradition, the human factor, and the place of forestry in the national economy as well as in the economy of average owners of small forest holdings will govern to a considerable extent the possibility of successfully transplanting proven institutions from the forest economy of one nation to another. This is a sound approach; hence it is important to consider how Sweden and the United States appear to differ in these regards. It is particularly important, I believe, because, unlike Dean Streyffert's paper, much that has been written about forestry in the two countries stresses similarities.

It seems obvious that forestry assumes greater importance in the economy of Sweden and in the economy of Swedish landowners than is the case in the United States. The value of forest products exported from Sweden annually, for example, is reported to vary from one-third to one-half of the value of all exports; whereas the United States is one of the world's leading importers of such products. Forest-based employment in Sweden also is said to be exceeded only by that in agriculture; but in the United States the per cent of the total labor force engaged in forestry and logging runs only about one-third of one per cent nationally (0.36 per cent in 1950), the maximum —about one per cent—is reached in only one forest region.

[1] W. B. Greeley, *Forest Policy*, McGraw-Hill Book Company, Inc., New York, 1953, p. 91.

The forest as a source of income to individual owners seems far more important in Sweden than in the United States. It has been authoritatively stated that the forest, including off-farm employment, is often the major source of cash income to Swedish farmers. On the other hand, the low incomes invariably attributed to their woodlots by American farmers in the Census of Agriculture has long been a matter of common knowledge and concern in forestry circles. Preliminary results of our recent work in North Carolina tend to substantiate the Census of Agriculture; among those interviewed, we seldom encountered a landowner, either farmer or other, who believed that over two per cent of his long-term average annual income could be attributed to his forest holdings. Woodlots also appear to offer a less significant source of home-use material to American farmers than is the case with Swedish farmers; but this is difficult to demonstrate because only fragmentary evidence exists on the subject. However, if I might again cite our recent work in North Carolina, we were quite surprised to learn that a very high percentage of the farm woodland owners in that state still made a regular practice of cutting timber for home-use purposes other than for use as fuel.

Still another important difference between American and Swedish forest conditions is in stability of ownership. American landowners are accustomed to an environment in which land is a more marketable commodity than it is in almost any place else in the world. And my own experiences indicate that they are very fearful of entangling arrangements which would restrict the marketability of their forest properties. These points can be illustrated by the fact that among nonindustrial private owners in the United States, the average length of forest land tenure ranges from about 10 to 15 years. Unfortunately, we have no reliable indication of the stability of family forest ownership which, incidentally, I found in Michigan to be correlated with the quality of forest management practiced.

I believe that there are also important differences with respect to Dean Streyffert's so-called "human factor and tradition" between American forest landowners and their Swedish counterparts. These are, of course, more difficult to quantify than the more tangible contentions; nevertheless, I am quite convinced that there are real and significant differences. American landowners have usually been characterized as rugged individualists. However, according to Schickele they seem to be "growing up to maturity from" what he characterizes as "the ruggedly egocentric stage of a spoiled child to the responsible stage of an adult who realizes that in order to enjoy freedom and personal dignity he must respect those precious values in others."[2] Swedish landowners, farmers in particular, must have begun to undergo such a transition nearly two hundred years ago, judging from Dean Streyffert's references to the breaking up of the old village communities and the consolidation of farms. The observations of several American writers on Swedish land policy would tend to substantiate these differences.

There also appear to be differences between Swedish and American forestry in terms of the advantages and disadvantages of small forest ownership. For example, forests both on and off the farm seem to offer little appeal to American farmers as a means of keeping their resources of machinery and labor fully employed. As a matter of fact, it seems that full-time farmers in the United States are tending to place increasingly higher alternative values on leisure at the sacrifice of part-time earnings. (I have been careful here to refer specifically to full-time farmers in order to distinguish that group from the relatively new, large, and rapidly developing class of part-time farmers in the United States, who are interested in agriculture only as a secondary source of employment.)

Another apparent distinction between Swedish and American private forest landowners is the relatively greater heterogeneity of that class of owners in America. Most Swedish forest owners appear to be farmers, while in this country, a large portion of the private forest ownerships and acre-

[2] Rainer Schickele, *Agricultural Policy, Farm Programs and National Welfare*, McGraw-Hill Book Company, Inc., New York, 1954, p. 5.

age is held by a conglomeration of other occupational groups. This heterogeneous character of American forest landowners, greatly complicates the problem of devising programs to encourage better forest management.

Dean Streyffert gives much attention to the relationship between size of a forest holding and its productivity in terms of economic efficiency. Such attention is well deserved, for it is basic to the subject under discussion. However, I fear that some of his statements (along with a great deal that has been said by Americans on the subject) could be misleading. Streyffert points out, for example, that the influence of the size of holding on the forest yield and on the economic outcome of forestry cannot be distinguished from the influence of ownership. Admittedly, size of holding is confounded with many factors of ownership which would make a straightforward empirical separation extremely difficult. However, modern research methodology makes it possible to consider the influence of size separately from that of ownership. The technique, which has been used in agricultural production economics, involves the synthesis of model farms representing various scale and enterprise-combination situations. This makes it possible to observe the changes in output which would result from adjustments in input levels, including differential adjustments among different enterprises.

We have not tried hard enough in the United States to apply these techniques of agricultural economics to our research into the economic problems of the small forest ownership. It is true that we lack refined physical or biological input-output coefficients for most forest practices; nevertheless, I believe that a great deal could be learned about the economics of integrating forestry and agricultural enterprises on individual farms by the employment of very crude or approximate input-output factors.

Finally, I should like to take partial issue with Dean Streyffert's summation of the interplay between size and efficiency of management on the one hand, and ownership and owner policy on the other. I refer to his thesis that ownership decides the policy of individual owners. It is my belief that it is better to strive to analyze the problem in a neutral context. Hence, I prefer to think that size of ownership and certain other characteristics of the owner are correlated with his total asset structure and his personal knowledge situation, both of which are large factors in determining the owner's forest policy. Economists commonly recognize this function of capital rationing and extent of knowledge with respect to the economic behavior of individuals. Incidentally, this suggests one way in which the orientation of our research into the entire small forest ownership problem could be improved. It seems quite fundamental to me that we should strive to learn more about the role of capital limitation versus that of knowledge deficiencies as they relate to this entire problem.

THE DEGREE TO WHICH CO-OPERATIVES HAVE BEEN DEVELOPED is the most distinctive feature of Swedish forest policy as it pertains to the problem of the small forest ownership. In his book on forest policy, Greeley states that "cooperation is the heart of Swedish forestry."[3] Dean Streyffert has brought out the very significant role that Forest Owners' Associations and Management Subsidiaries play in Swedish forest policy. As nearly as I can deduce, the difference in the two types of organization is largely one of degree. Management Subsidiaries appear to serve both production and marketing functions more or less within the larger Forest Owners' Associations but more intensively than associations. For my purpose here the distinction does not seem great enough to warrant discussing the Associations and Management Subsidiaries separately, so I shall refer to them as if they were almost synonymous terms.

It has been pointed out by Dean Streyffert that farmers' co-operation was a necessary forerunner of the progress that has been made through forest co-operatives in Sweden. It thus seems advisable to very briefly consider the history and importance of agricultural co-operatives in Sweden and in this

[3] W. B. Greeley, *op. cit.*, p. 93.

country. According to Marsh,[4] in 1954 co-operatives handled nine-tenths of the salable agricultural produce of Sweden. Membership in co-operatives is quite large and most Swedish farmers appear to belong to more than one, hence total membership figures do not reflect total proportionate participation. However, it seems significant that membership in nearly all types of agricultural co-operatives in Sweden has had a continuous history of growth.

Co-operatives have been far more successful in American agriculture than in any other sector of our economy. The number of agricultural co-operatives in the United States reached a peak in 1923 though total membership and the dollar value of transactions have continued their steady rise. In the American agricultural economy, co-operatives tend to be concentrated geographically and in terms of classes of crops or products. For example, approximately 60 per cent of the total membership in farm marketing co-operatives is concentrated in twelve states in the North Central Region, while about 42 per cent of the total American membership is among co-operatives dealing in two classes of commodities—fruits and vegetables and dairy products. It seems significant that this geographical concentration of participation in co-operatives among American farmers is found in that part of the nation where the farm population is made up of a high concentration of people of Scandinavian origin.

There seems little doubt that co-operatives offer some possibility toward a partial solution of American problems of small forest ownerships. In seeking out these possibilities, I am sure that much can be learned from experiences in agriculture and in Swedish forestry. However, I believe that many of the writings on the subject are quite misleading with respect to the matter of marketing efficiencies. The misleading statements appear to have arisen as a result of the superficial treatment given the subject or of implicit assumptions that were

[4] Raymond E. Marsh, *Public Policy Toward Private Forest Land in Sweden, Norway and Finland,* Charles Lathrop Pack Forestry Foundation, Washington, D. C., 1954, p. 5.

fallacious. The type of statement which I am criticizing would lead us to believe that forest co-operatives or owner associations can automatically overcome the cost diseconomies of size involved in marketing forest products from small properties.

It does not seem to me that anything in the inherent efficiency of the co-operative system would necessarily make the costs of performing its marketing services less than those of any other system providing similar services. Nor is there any reason to assume that co-operative marketing would automatically provide a more efficient service for small scattered forest holdings which in turn would result in higher prices being received by owners. If cost reductions and higher returns to owners are effected through substituting a co-operative system for some other system of marketing, it is most likely due to one, or a combination, of four things.

> First, the bargaining power of the owner or seller has been improved relative to that of the buyer through a shift in their relative monopolistic positions.
>
> Second, some of the services of marketing and their associated costs which were formerly performed by the buyer have been shifted to the seller or owner; hence, the apparent increase in prices received by the owner was a return for the added services he provided himself.
>
> Third, actual cost economies of size have been realized by permitting more specialization in the services rendered as a result of having expanded the over-all size of the marketing operation, or by having substituted the specialized services of the co-operative for those of a less specialized nature.
>
> Fourth, reduced average unit costs may have been achieved by the simultaneous handling of several differentiated but related commodities which were formerly handled separately.

One or more of these changes must be involved with a shift to co-operative marketing before savings in the costs of marketing can arise. There is nothing inherent in the co-operative system itself that assures such changes.

Careful examination of the history of forest co-operatives

will tend to substantiate these arguments. Dean Streyffert alludes to several of these explanations for the success of forest co-operatives in his country, including for example, the implication of shifts in bargaining power, shifts in the person providing part of the services, increased specialization of labor, and the simultaneous handling of different forest products. Incidentally, the tendency toward a shifting of monopoly strength from the buyer to the seller following the establishment of forest marketing co-operatives was observed in the Norwegian forest economy by Zivnuska,[5] who also noted that buyers had tended to counter their relative loss of strength by resorting to more centralized action.

Success that Forest Owners' Associations and Management Subsidiaries have enjoyed in Sweden and elsewhere in Scandinavia as a solution to many of the problems of the small forest ownership, immediately suggests the question of their prospects in the United States. As a result of studies in American agriculture, farm economists have made some interesting observations with respect to the prerequisites for the successful founding and operation of co-operatives. Foremost among such prerequisites they mention that prospective participants must feel a definite need for some special effort to cope with a situation which is bothersome to them. Dean Streyffert and others who have written about forest owner associations in Sweden have indicated beyond a shadow of doubt that Swedish landowners felt a definite need for the services they could obtain through co-operative action. On the basis of my own experiences with American forest landowners and from what has been written on the subject, I sincerely doubt that a sufficiently strong feeling exists among them to offer much hope for the successful founding of co-operatives unless such an urgency could be stimulated through advertising techniques.

[5] John A. Zivnuska, *Private Forestry in Norway—A Case Study in Small Woodland Management and Policy,* Forest Science Monograph 1, Society of American Foresters, Washington, D. C., 1959, p. 32.

In Sweden one of the strongest stimuli for the establishment and expansion of forest co-operatives apparently has been in the area of timber marketing. Down to the present time, however, I do not believe that the average American forest owner has been greatly concerned about his marketing problems and for two reasons. First, most of the volume and value of timber products cut from small forest properties has gone into the sawtimber market which generally has been a relatively competitive market. Second, marketing studies have shown that the majority of owners making sales of sawtimber stumpage have been relatively satisfied in their dealings with local buyers. I can visualize, however, that as our forest marketing economy continues its present gradual shift away from sawtimber and toward the less competitive pulpwood buying market, a feeling of need could develop among owners for organizations to counter any monopolistic advantages displayed by buyers.

Another strong reason why Swedish forest owners felt a need for co-operatives appears to have been a result of labor shortages and the consequent difficulty in obtaining cultural and other needed services for their woodlands. I believe that the more progressive American forest owners now feel the need for similar services and that they are not being adequately provided. Our public service foresters and private consultants are convincing forest owners of the need for performing cultural work in their woodlands. However, a large fraction of this successful effort has been lost because forest owners with available investment capital have become discouraged in their efforts to purchase the services required to bridge the gap between recommended and applied practices.

Another prerequisite to the successful operation of a co-operative type venture is ability to distinguish between belief and reality. Many co-operatives launched without prior investigation of whether there are substantial profit margins, a need for services not presently available, and other indications of an opportunity for a profitable operation, have resulted in failure. Co-operatives seem extremely vulnerable to low volumes of business inasmuch as success depends to a considerable extent on being able to operate on a narrow margin. Insufficient volume of business has been a large factor in the failure of several ventures into forestry by successful agricultural co-operatives. A few years ago we had the opportunity in North Carolina to observe the disappointment sustained by a very successful agricultural co-operative in its forestry venture because of the low volume of forest products channeled through its organization.

Two important points can be singled out as those most often overlooked by persons who have suggested co-operatives or owner associations as a panacea for the problem of the small forest ownership in the United States. The first concerns the apparent willingness of people to join in co-operative action to solve economic problems, and relates to Dean Streyffert's point concerning the role of "tradition" or the "human factor." We have already noted the unequal geographical distribution of co-operatives in American agriculture; it seems safe to regard this pattern as a good regional indicator of the relative acceptability of co-operatives to landowners. If this assumption holds, it follows that landowners are least prone toward co-operative action in several sections of the country where we recognize the small forest ownership problem to be most serious. And it also happens that these areas are coincident with two of our most productive commercial forest regions—namely New England and the South.

The second point, seemingly given little consideration by the enthusiastic supporters of owner associations has to do with the difficulty of initiating such a program even if it is clearly deemed advisable. Those who have analyzed causes of success and failure among American co-operatives have repeatedly pointed to the problem of attracting and retaining capable management amidst the competition from other types of business organizations. It seems to follow from this that even stronger leadership than would be acceptable at a later stage would be required initially to organize a co-operative or owner association and to maintain interest in it through its early stages of development.

IF THESE COMMENTS ON FOREST CO-OPERATIVES HAVE CREATED AN impression that I believe they have little to offer as a solution to the small forest ownership problem, it is merely the result of my efforts to offer a critical analysis. Actually, I am convinced that Forest Owners' Associations as well as other Swedish institutions of forest policy offer much potential to the solution of many of our private forest ownership problems providing we do not forget Dean Streyffert's advice on the necessity of exercising care in adapting such institutions to different environments.

Let us consider, for example, the last difficulty cited above concerning the initial organizing of Forest Owners' Associations. How might this be overcome within the framework of American institutions? The adaptation might be effected by drawing on three typically American institutions quite familiar in forestry: (1) the long-term forest management lease; (2) the independent private consulting forester; (3) the lending precedent of the Small Business Administration. It is quite conceivable to me that these three quantities could be blended with a minimum of legal reform to provide the leadership initiative needed to create and sustain an organization similar in function and advantage to Swedish Forest Owners' Associations. Given this suggested backing of low-cost long-term government loans, I believe many ambitious and enterprising firms of young and dedicated forestry consultants would rise to the challenge.

To return briefly to the subject of marketing, it seems to me that one common misconception is especially in need of clarification since marketing has become an increasingly popular subject among American foresters when they discuss the difficulties of managing the small forest holding. Nearly always, I am certain, these men have not been speaking of marketing efficiencies or other marketing topics commonly discussed by economists. Instead, they seem to have had in mind the need for creating or increasing the demand for forest products. And frequently these fuzzy analyses have led them to suggest the formulation of co-operatives on the assumption that they could automatically solve such problems. Admittedly, co-operatives should have the incentive for attempting such a solution but chances are it would require a larger capitalization than normally envisioned for forest co-operatives.

Perhaps one of the foremost lessons we could learn from Swedish forest policies lies in the realm of administrative organization and co-ordinated action. In describing Swedish forest policy and administration Greeley has stated that "the semiprivate forest associations and cooperative societies bewilder and impress the American student by their multiplicity, by their quasi-official status, and by the effective ways, through law or custom, in which their functions integrate in getting the things done that are necessary to the successful practice of forestry."[6] By comparison, our own multiple forestry programs seem simple; yet many students of forest policy have criticised our inability to co-ordinate their functions. This sad state of affairs has certainly not resulted in the maximum assistance being rendered to owners of small forest properties and has prompted Gulick to describe the situation as one of "confusion at the grass roots."[7]

This latter point concerning the administration of United States forest policy is just one of many that could be mentioned as suggestive of the need for objective and impartial research into our entire forest policy structure. Such studies, if undertaken, should begin with the basic underlying assumptions for I am sure that in the past we have been guilty of predicating entire programs on faulty or unrealistic assumptions.

In conclusion, let me at least mention another fundamental factor in any solution that may be worked out—the characteristics of the owners themselves. In the light of my contacts with unbiased samples of forest landowners in widely separated parts of the country, I am constantly amazed at even the limited progress that has been made thus far. Most owners work under

[6] W. B. Greeley, *op. cit.*, p. 91.

[7] Luther Halsey Gulick, *American Forest Policy,* Institute of Public Administration, Duell, Sloan and Pearce, New York, 1951, p. 184.

heavy handicaps. Indeed, I sometimes think that forest landownership must be the heritage of the underprivileged; for many of our small forest properties are in the hands of elderly people, partially physically incapacitated and extremely limited on capital.

On the brighter side, I have always been encouraged by the contrast the younger forest owners present. They are much better informed and are generally anxious to adopt new technology, though often lacking the capital to do so. This leads me to suggest that perhaps the most rewarding and least revolutionary institutional change possible for coping with this small ownership problem would be one to facilitate the transfer of forest properties to younger owners without heavily burdening them with debt in the process.

III GREAT BRITAIN:

PRESERVATION OF NATURAL AREAS

STANLEY A. CAIN

The Nature of the U. S. Problem

E. M. NICHOLSON

● *Preservation of Natural Areas in Great Britain*

EDWARD H. GRAHAM

What the United States Can Learn

1. BEINN EIGHE NATURE RESERVE, ROSS-SHIRE. The eastern slopes of Beinn Eighe, with remnants of the Old Caledonian Pine Forest.

2. COED TREMADOC NATURE RESERVE, CAERNS. On the steep cliffs in the foreground is the remnant of the oak woodlands that once covered most of Snowdonia.

British Crown Copyright

3. BRIDGWATER BAY NATURE RESERVE, SOMERSET. Stabilization of mud on banks of the River Parrett by planting of *Spartina townsendii*.

4. BELOW: Left—ROUDSEA WOOD NATURE RESERVE, LANCASHIRE. Caterpillars dislodged from an oak by an injection of insecticide are swept up in a population study of *Tortrix viridana*.

5. Right—MOOR HOUSE NATURE RESERVE, WESTMORLAND. V-notch weir for measuring water runoff.

Both photographs British Crown Copyright

6. WOODWALTON FEN NATURE RESERVE, HUNTINGDONSHIRE. The "island" of the reserve shows scrub growth on unreclaimed fenland. Famous as the home of the Large Copper Butterfly—*Lycaena dispar batavus.*

7. YARNER WOOD NATURE RESERVE, DEVON. The reserve is in the middle distance and the right foreground. In the middistance at the right are experimental treeplanting plots. The edge of Dartmoor is in the background.

THE NATURE OF THE U. S. PROBLEM

■ STANLEY A. CAIN

IN THIS INTRODUCTORY PAPER I shall try to highlight the advantages and shortcomings of our system of natural areas in the United States in a way that will provide a domestic background for the essay that follows: E. M. Nicholson's discussion of problems and programs of preserving natural areas in Great Britain.

First, what is a natural area? Since our approach is essentially practical rather than philosophical, let us avoid overstrict definitions. All cats are gray in the darkness of our present knowledge. The natural areas with which we are concerned here have only degrees of naturalness which are related to the intensity of past and present influences of man on nature. For example, although the vast tropical rain forest of the Amazonian hilea apparently heals itself rapidly under shifting

■

STANLEY A. CAIN is Charles Lathrop Pack Professor of Conservation and chairman (from 1950 to 1961) of the Conservation Department of the School of National Resources of the University of Michigan. He also is professor of botany in the University's College of Literature, Science, and the Arts. He is a member of the National Parks Advisory Board and of the Commission of the Michigan Department of Conservation. During his long career as a teacher and researcher Mr. Cain has written widely in the fields of botany and conservation. He was born in Jefferson County, Indiana, in 1902. He received his B.S. degree from Butler University, Indianapolis, and his Ph.D. from the University of Chicago.

agriculture, those most familiar with local regions in the Basin claim that disturbances by pre-Columbian Indians still can be detected after several centuries. In another frame of time reference, I have not hesitated to consider that certain tracts of forest that I have studied in Tennessee, Indiana, and Michigan were, for most scientific and academic purposes, primeval, even though I knew or suspected that they had been "high-graded" for black walnut or black cherry logs a century or more ago. Under some circumstances, an impounded area of marsh and swamp may develop a naturalness with the passage of time that is entirely comparable to wet areas caused by beaver or geomorphological changes that embarrass drainage, even though man may have built a dam.

All this suggests that it would be practical and even reasonable to have a sliding scale of values with regard to the naturalness of natural areas. For example, two aspects might change directly with the distance of the area from centers of dense human population: the closer it lay to human aggregations, the smaller it could be without losing its interest, and the greater could be its tolerable degree of departure from complete naturalness. Because the degree of naturalness of the landscape tends to increase with isolation, the best protection afforded a natural area is its inaccessibility. We have here a significant comment on the building of roads, trails, and other facilities for human convenience and the problem of protection of nature.

IN THIS COUNTRY NATURAL AREAS SOMETIMES ARE CRITICIZED AS single-use areas, and such criticism often is followed by a stout statement of belief in multiple-use as good conservation. Persons taking this tack usually are interested in promoting for profit some incompatible use, such as mining, oil and gas production, timber harvest, hunting, or tourism and its related commercial developments. Granted that as a nation we have become a net importer of raw materials, including forest products, certain metals, and petroleum, is our need so great that we must sacrifice to it the less tangible uses of natural areas? What are some of these values that do not result in marketable, movable commodities?

One value of nature is expressed in *Goals for Americans,* the recent report of the President's Commission on National Goals. In its general conclusion, the Commission says, "The very deepest goals for Americans relate to the spiritual health of our people. . . . Indifference to . . . values other than material comfort and national power . . . is inexcusable in a society dedicated to the dignity of the individual." I take it that it does not require proof that the human spirit may be refreshed by experience with nature.

Another line of reasoning was brought forward recently by J. Donald Adams in *The New York Times.* He wrote, "The more man becomes herded in great agglomerations, in which, through technological advances, he becomes more and more divorced from the natural world, as simpler, more rural civilizations were not, the more even his agriculture becomes mechanized and scientifically administered, he loses touch with the other forms of life to which he is bound by ties stronger than he has become able to realize. His arrogance must make way for a new humility." These words were written in a discussion of the ecological aspects of nature, the beauties and lessons to be found in the "balance of nature," and the hazards to be found in the "most misleading slogan of our age—the conquest of nature." Man has so much yet to learn from nature that his extirpation of any type of natural area from the face of the earth would be at the risk of his own peril.

The greatest weakness of the American system of natural areas, I think, is that we lack a broadly based, well financed program of ecological investigation of natural areas. We need such a program because it is basic to adequate interpretation to the public of nature and the significance of natural areas. A stronger interpretation program, based on research, would not only enhance the pleasure of visitors to National Parks and other natural areas, but would also broaden the base of public support for the perpetuation of natural areas when conflicting uses seek to pervert them to their own ends. In the second place, the fruits of a thoroughgoing ecological research program on natural areas would, I am sure, provide invaluable suggestions for better management of other lands for more commonplace, materialistic ends.

However, during the past fifty years we have so emphasized—perhaps overemphasized—protection of nature in National and State Parks that manipulative research is precluded under present policies. Perhaps, to avoid weakening the policy of protection, we should seek, as has the Michigan Natural Areas Council, the establishment of "natural research areas," to join its sister categories of "nature study areas," "natural area preserves," "natural area reserves," and "natural scenic sites."

THE NATIONAL PARKS FORM THE FOUNDATION OF THE AMERICAN system of natural areas. The special Act of Congress establishing Yellowstone National Park in 1872 provided "for the preservation, from injury or spoliation, of all timber, mineral deposits, natural curiosities, or wonders within said park, and their retention in their natural conditions." It also directed that Yellowstone be "a public park or pleasuring ground for the benefit and enjoyment of the people."

Ever since the National Park Service was created in 1916, successive administrations have contended with these two sometimes conflicting objectives—preservation of nature and public enjoyment of it. Constantly under opposing public pressures, on one side for increases in facilities for enjoyment of the parks by rapidly growing numbers of vacationists and on the other side by defenders of the integrity of the wilderness, the Service has in general done a highly commendable job of preserving perhaps 95 per cent of National Park acreage from development, while fostering within the limits of its budgets the means of comfortable enjoyment of the Parks by many millions of citizens each year.

In 1906 the Antiquities Act gave the President authority to create National Monuments from suitable lands in the public domain. In addition to historic and prehistoric remains, the Monuments may include landscape of especial interest because of geological, botanical, or wildlife features. Such natural areas, which generally are of smaller acreage than the National Parks, are administered under the same policy of protection and public enjoyment.

A constantly alert Service and public are both necessary to protect the integrity of the National Parks and Monuments and to preserve their natural areas from encroachment and deterioration. Other public agencies, both federal and state, with special responsibilities and objectives often find their development plans in conflict with the preservation policies of the Park Service. Notable cases have occurred in connection with the direct need of municipalities, industries, and agriculture for water, and the general need for hydroelectric power. Congress itself has produced some confusing cases.

Within some of the National Parks and Monuments private commercial interests hold some rights that are incompatible with the basic objectives for which the public reservations were established. Also difficulties are often caused by about three-quarters of a million acres of intermingled private land inholdings. Basically, these public natural areas exist to provide for certain nonconsumptive, nondepleting services, yet there is persistent demand by local pressure groups and special-interest associations for commercial use of the parks for mining, grazing, timber harvesting, hunting, and the production of other economic goods.

A different difficulty arises from local demand for the inclusion in the System of areas of substandard quality, in the national sense, because of commercial benefits that would result from increased local tourism.

Perhaps the most insidious pressures arise from the legitimate desire of the public to enjoy what are, after all, their own National Parks. These pressures in the first place arise innocently from mere numbers of visitors and the service needs which they have. Park administrators have to struggle with their desire to satisfy adequately the public need for campgrounds, cabins, hotels, sanitary facilities, and roads and trails, and their urge to "show off" their parks to the best advantage to the largest number of people. It is, I suppose, no more than human, or perhaps American, to take pride in size, numbers, completeness, and coverage. As Marion Clawson has pointed out, "over-accessibility" is a subtle and dangerous threat to the natural features of every park.

I am convinced that an interest in nature is the deepest motivation of park visitors although in most of them it may be quite unsophisticated, and in many deeply submerged below conscious levels. As regards the wilderness, most park visitors are satisfied to be able to see from vantage points, and in most cases from their own automobiles, vast stretches of undisturbed landscape, and only a small number seeks intimate contact with nature. Yet there is an unthinking pressure on the part of much of the public for more roads and trails everywhere, airplane landing fields, and marinas for the enjoyment of high-speed boating. It is thoughtlessness and ignorance of the basic purposes of National Parks and Monuments on the part of a segment of the public that lend strength to the pressures, abetted by the interests of certain kinds of private enterprise, for golf courses, ski tows and related facilities, and other "improvements," not excluding amplified music and even lectures and the illumination at night of outstanding features.

The conclusion is inescapable that the integrity of the Parks and Monuments can be maintained only by the unobtrusive location of all public service features and the rigorous exclusion of all facilities for organized sports and mechanized pleasures.

Among other kinds of Federal land classifications that include natural areas, are the migratory waterfowl, general wildlife, big game, and other National Wildlife Refuges. It is not necessary here to say more about them than that they permit certain kinds of management for specific ends that are outside Park policy. At the same time, it is important to recognize that within the Parks it is necessary sometimes to undertake measures for disease and pest control, for animal population control, and for fire control that are in a strict sense a departure from complete naturalness.

It is important, however, to say a word about the system of wilderness and wild areas that has been developed by the Forest Service in the Department of Agriculture. In general they are comparatively large areas of high scenic value and either low timber-producing capability or present inaccessibility. They do not have the "security" of law that natural areas in the National Park System have, and having been established by agency administrative order can be opened at any time for timber harvest, mining, or other product development or commercial services.

In the near future we can expect considerable national debate concerning the transfer of Forest Service lands or of the responsibility for their management from one federal agency to another. There are thoughtful commentators who believe that the national purpose would be served better by the transfer of certain park-quality lands from the Forest Service or other agencies to the Park Service. Others, backing the multiple-purpose concept of land management, see no reason why the Forest Service cannot adequately do its share toward satisfying the national need for natural areas. Still others feel that the National Park Service job has been made impossibly difficult and the security of the natural features of Parks diminished by the confusion of activities that have resulted from the Park Service's responsibility for recreation on Bureau of Reclamation's reservoirs and the new drive for National Shoreline Recreation Areas. Some of these persons believe that a new National Recreation Agency, which would relieve both the National Park Service and the Forest Service of all but extensive "wildlands" recreation, would be a step in the right direction. In any case the multiplication of National Park Service responsibilities, as the variety of areas and activities in its charge has increased, and the current drive in the Forest Service to make multiple-use its guiding policy as exemplified by Operation Outdoors, have brought both agencies deeply into the recreation field in a way that may be inimical to the full protection of natural areas.

Park systems in the several states and units of local government are very widespread. In passing, it can be noted that a further objective of the National Park Service is "to provide assistance to other Federal agencies and to the States and their political subdivisions in the planning of public park and recreation area facilities." Certain proposals before Congress if enacted will provide Federal matching funds for the acquisition of land for state and local parks and recreation areas.

Without attempting detailed consideration of the nature, extent, protection, and use of natural areas that exist in these nonfederal units it can be said that nature protection at the state or local level is commonly more difficult than it is at the federal level. This is largely because the legislatures, local governments, and land-administering agencies are more sensitive to economic needs and more subject to strong local interest groups.

In addition to public efforts, there are individuals and private groups who own land for the purpose of preserving natural areas. Experience with individual holdings suggests that in general the heirs of such persons are more interested in liquidating an estate than they are in perpetuating their ancestors' eccentricity and that in consequence the virgin timber is cut, the land goes to a developer, or some similar fate occurs. On the other hand, the Audubon Society and the (United States) Nature Conservancy are examples of citizens' groups that are deeply interested in the preservation of natural areas, including their ownership. Although important in the total picture, the capacity of private interests to preserve natural areas is essentially limited, local, and particular. The principal responsibility would seem to be a public one.

HOW TO KEEP A NATURAL AREA NATURAL AFTER IT HAS BEEN SET aside is a problem that continually plagues agency administrators and the interested public. The act of allocating a tract of land to noncommercial uses and dedicating it as a park, wilderness, or other natural area does not warrant a reduction of vigilance on the part of those who protect it from competing uses.

Let me illustrate this general observation with a few recent examples from the State of Michigan, reflecting situations with which I am well acquainted. While it is true, as we have just noted, that pressures within states are commonly more intense than at the national level, the basic issues and the nature of the conflicting interests are quite similar.

In 1958 a mining exploration company applied to the Michigan Department of Conservation for a permit to explore a copper-bearing formation known to have a small outcrop on the shore of Lake Superior at a location well within the boundary of the Porcupine Mountains State Park. The Park had been established by an Act of the Legislature in order to preserve a scenic remnant of wildlands clothed with primeval forest of the Great Lakes Northern Hardwoods type. The issue was clear: were the interests of the State better served by preservation from development of this last large remnant of original Michigan landscape, or by the increased economic activity for a long-depressed region that would result if the exploration were to lead to mining development? Individuals and organizations quickly lined up on either side of the question. The Interim Committee of the Legislature and the Michigan Department of Conservation both held public hearings, and the Department's Commission postponed action on the mining permit for several months until it became clear that the mass of public opinion was for preservation of the wilderness qualities of the Park. The Department's Geology Division, which seemed originally to have favored mining, ultimately wrote a set of requirements under which mining could proceed (in the event the Commission approved mining) that were so strict that a mining operation could scarcely have been profitable. The mining company withdrew its request and the Commission never had to vote on the issue.

The Porcupine Mountains State Park, with its more than 50,000 acres of unique wilderness, does not remain secure, however, even with the protection of its special legislation. In the 1961 *Timber Producer,* the executive secretary of the Timber Producers Association of Michigan and Wisconsin started a full-scale attack on the Department of Conservation with a view to promoting timber cutting in the Park. He charged that garden clubs, bird watchers, sportsmen's clubs, and park and natural-area protectionists are not true conservationists, and "it is a false premise to assume that conservation to the point of rot, waste and loss [of timber] is good conservation whether in our state parks or any other area of natural resources."

In 1960 a lumber company petitioned the Department of Conservation for the timber on 180 acres in Tahquamenon Falls State Park. This small acreage, which carries virgin northern hardwoods forest and is near a timber-cutting operation on private land, is not only within the boundary of the Park, it is also within a Natural Area Preserve. The latter classification "guarantees" maximum protection from all uses which would affect an area's naturalness. Nevertheless, serious consideration was given to the proposal by the Department of Conservation because the money that would have been received for the stumpage would have permitted acquisition of a considerable acreage of private inholding within the park.

There is a considerable flurry in oil and gas prospecting and development in Michigan. During 1960 the Department of Conservation leased 203,862 acres; receipts from these leases amounted to more than $780,000. Of the 95,000 acres in state parks or recreation areas, more than one-third are currently under oil or gas lease. Leases on these two types of land include a special provision that drilling permission must be granted as a separate act by the Conservation Commission. Recently the first request for such permission to drill a wildcat well was rejected by a one-vote majority. (It seems probable that the permit would have been granted by a one-vote majority had the full Commission been present.) When the oil company contested the Commission action on the point that selling the lease implied the right to drill, the State Attorney General gave an opinion that the Commission had clear power to act as it did. The Commission has since adopted a new firm policy statement which precludes drilling on park and recreation lands except to avoid drainage by nearby wells on private lands. However, the case does not seem to be closed; the oil and gas companies are prepared to resist any threat to what they consider their right to produce oil and gas wherever they may occur, including parks and other natural areas. The State Commission now has three new members and its decision on such an issue is not known.

In Michigan as nearly everywhere else in the United States, the building of new roads and highways and the improvement and rerouting of old ones present a continuing threat of encroachment on natural areas. Because of the strong public support for good roads and the tremendous funds currently available for construction, highway departments tend to ride roughshod over all other land uses. There is considerable evidence that highway routes are sought over public lands wherever possible. And why not? Easements are either free or cheap to buy; a public agency is easier to deal with than a large number of disgruntled private owners; routes often are more scenic than they would be otherwise; and in any case wildlands are frequently considered by highway personnel to be "waste" lands.

These few examples are sufficient to show the character of the external pressures on agencies that administer public natural areas. Even when areas are protected by specific law, there is a continual effort to remove such protection in the interest of economic development. When natural areas are protected only by administrative policy within broad legislation, they are subject to shifts of policy that occur with changing pressures, especially if there are no explicit policy statements and no unanimity of opinion within the agency.

When a state has been under an austerity program for several years and the available funds are inadequate to maintain, much less expand, the acreage and facilities to meet the growing pressures of public land uses, it is very easy for administrators to look with favor on those land uses which promise to return significant dollars directly to the agency or the state. Lands held as natural areas provide only public services. These services are less tangible than the goods which could be gained by development. It takes a wise and tough-fibered administrator to equate the values of a wildlife area (especially of non-game species) with barrels of oil, or those of virgin forest with pounds of copper or sheets of birch veneer.

Not all encroachments on natural areas result from external pressures for conflicting land uses. Sometimes within agencies administering natural areas, personnel with different specific responsibilities may cause a loss of or a deterioration in quality of natural areas. For example, those who design the plans for a park may locate roads, trails, impoundments, parking lots, camp grounds, service buildings, concessions, and other public facilities with little or no regard for the values which are implicit in natural areas. There is, of course, less excuse for this kind of loss of natural areas than there is in the case of external pressures on an agency that are brought by private enterprises.

TO SUM UP, OUR SYSTEM OF NATURAL AREAS IN THE UNITED STATES is extensive in comparison with those of Great Britain and other older, more densely populated countries. And, although "wildness" is a relative quality, much of the American acreage still is comparatively unspoiled. Federally held lands, the National Parks and Monuments in particular, contain the largest and best protected remnants of the original landscape. The several State Park systems allow for the preservation of smaller areas and for completing the array of national landscape types. Some additional protected and undeveloped areas are maintained by private individuals and organizations. However, the principal responsibility is, and, I think, should be a public one.

But there are still gaps in the system; the greatest need being for comprehensive programs of ecological research. Also, few states have a systematic program of preservation of natural areas for the purpose of completing, as outdoor museums, the range of types found within their borders. Far more serious, however, is the rapid encroachment on wild lands of many kinds of development, even in remote places.

The acquisition of title to appropriate lands, and their dedication to the compatible uses of natural areas, remains a pressing problem.

The battle for the preservation of nature and the exclusion from natural areas of inappropriate uses is never completely won. It has to be fought again and again as new threats to the integrity of natural areas arise or old ones are renewed.

PRESERVATION OF NATURAL AREAS IN GREAT BRITAIN

■ E. M. NICHOLSON

AFTER THE LAST GLACIATION Great Britain, like North America, became largely tree-covered. Forest clearance in Britain, however, took place in the main between half-a-dozen and a dozen centuries earlier, and has for the past five hundred years been far more nearly complete and final than even to-day in many parts of America. Land uses in succession to primeval forest therefore have a much longer history, a more complex character, and a more varied evolution in Britain.

Indeed, almost everything about the problem is bewilderingly complex. Geographically Great Britain is an island nearly the same size as the state of Oregon and with a some-

■

EDWARD MAX NICHOLSON is Director-General of the Nature Conservancy in Great Britain, and Secretary of the Privy Council Committee for Nature Conservation. He was a charter member of the Nature Conservancy from its foundation in 1949, and a member of the Wild Life Conservation Special Committee 1945-47 which drew up the British official scheme for wildlife conservation. He has played a prominent part in the International Union for the Conservation of Nature and Natural Resources; has undertaken missions of advice on the handling of conservation problems for the governments of Ontario and of Northern Ireland; and has served as officer, council member, or trustee of numerous other organizations in the conservation field. Mr. Nicholson has also been active in other fields, as a member of the Advisory Council on Scientific Policy and as a Trustee and former General Secretary of P E P (Political and Economic Planning). He was born in 1904 and educated at Sedbergh School and at Hertford College, Oxford.

what similar climate, but lying so much more northerly that if it were shifted westwards to this longitude it would straddle most of the eastern shore of Hudson Bay. In certain spots palm trees, bamboos and fuchsias flourish in the open up to about latitude 58, nearly 20 degrees north of Washington, while close at hand salt-laden high winds forbid any trees to grow, and at best the climatic tree-line runs below 2000 feet above sea level. On some coasts the annual average wind speed is over 12 miles per hour. (Stornoway had 51 days in 1959 when the mean hourly wind speed was 39 miles per hour, or more.) Annual rainfalls range from under 20 inches to over 150 within comparatively short distances. Geology is fantastically complicated and there is a wide range of variety in soils, vegetation, and animal population. Even among the human inhabitants, at least a dozen major types are so different in outlook, customs and sometimes language and speech as to have to be taken into account in good administration. It is therefore almost impossible to frame a generalization about Great Britain (or even about the component nations, England, Scotland and Wales) that is not subject to more exceptions and qualifications than are manageable in a paper of this compass, which can only be based on a personal judgment of what are the most significant problems, factors, and trends.

Thus the British problem of preservation of natural areas has arisen in a very different geographical, biological and historical setting, and has consequently emerged in a different shape, and become viewed from a very different standpoint, than the North American. Much of what is plainly visible here in the impact of man on primitive nature is lost in the mists of history or legend. Americans still living can perhaps remember people who were contemporaries of Davy Crockett; our frontier days are farther back, and we cannot even be sure which bird was the "woodwele" which sang,

> and wolde not cease
> Sitting upon the spray,
> So lowde he wakened Robin Hood
> In the greenwood where he lay

Already by A.D. 675 persecution of wildlife had reached a pitch which led St. Cuthbert to establish on his Farne Island retreat off the Northumberland coast the oldest of European sanctuaries, which is maintained with loving vigilance at this day. Loss of wilderness was one of the urgent problems which faced King William the Conqueror when he inaugurated his new administration in 1066. He anticipated the American National Parks by establishing a series of Royal Forests to conserve some of the last substantial remnants of English wilderness. These Royal Forests were closely and accurately demarcated, and on the whole vigilantly and efficiently administered, under a code of law and management designed to conserve not only the deer and other wildlife but the habitat, or "vert" as it was called. Some of the penalties for infringement have been modernized, but, in the 65,000 acres of the New Forest, Crown management for conservation has been continuous over nearly nine centuries, and during the past two years the Nature Conservancy, by agreement with the Forestry Commission, have undertaken nature protection duties in this ancient reserve. It gave us great pleasure last June that several leading American conservationists were able to walk through it with a number of ours in commemoration of the historic walk along the same route fifty years earlier by Theodore Roosevelt and Edward Grey, just at the time when the great American conservation movement was springing to fame.

Unfortunately, apart from the New Forest and Wychwood Forest—now a National Nature Reserve—nearly all our original series of officially conserved national areas have been filched away and destroyed by those encroaching interests which, like jackals and hyenas, are always hanging round in the shadows waiting their chance to supersede trusteeship by exploitation.

In passing on from the days of falconry and the royal chase it should be noted that the ideas of what we should now term game-cropping, which underlay the medieval management, are in the past three years regaining a key position in some of our latest management plans, and even more in the management of wildlife in Africa. Without the powerful tradition of field sports and game preservation, continuing in different forms until the present day, a large part of our surviving remnants of natural and seminatural habitats in Britain would have been lost. The pervasive and enduring influence of this interest, sometimes wrong-headed and destructive, but on the whole farsightedly conservationist, is one of the factors which must always be borne in mind in any comparative judgments on British and American conservation policies, practices and achievements. Something over a third of the British National Nature Reserves and well over half their acreage are either held by lease or agreement on sporting estates or were up to their recent purchase by the Nature Conservancy managed primarily for sporting purposes.

The other great conservation problem with a long history and an enduring influence in Britain is that of trees. Recent research has shown that some of our most interesting aquatic nature reserves in the Broads of Norfolk are not, as was until very recently believed, of natural origin but were the result of failure by the dense early medieval population of this area to conserve their woodlands, so that they were forced to find a substitute fuel by undertaking vast and deep excavations for peat, which became flooded from about 1350 onwards, and were thus transformed into the most extensive and interesting system of inland waters in England outside the Lake District.

Shortages of fuel were accompanied by shortages of wood as material. An acute need for conserving oaks was felt by the Navy, especially in the seventeenth and eighteenth centuries and this led (through the influence especially of John Evelyn) to strong and widespread enthusiasm for tree-planting, and for the introduction of exotic species, which has left a massive mark on British vegetation and landscape. This fashion, unparalleled in any other country, has bequeathed us not only much picturesque scenery but also many problems in the management of artificially modified woodland nature reserves.

Already in the sixteenth century the hideous problem of the impact of technology on natural areas was posed by the growth of charcoal-using blast furnaces, the building of which had to be banned in some areas on conservation grounds, so that ironmasters were forced to migrate to more remote sites. Eventually in 1709 one of them, Abraham Darby achieved at Coalbrookdale in Shropshire one of the decisive breakthroughs of industrial history by managing to use coke instead of charcoal as a fuel for smelting iron, and thus emancipating his industry from shortages of wood and from the technical limits which it imposed. Despite numerous similar instances, industrialists and engineers two hundred and fifty years later are still more inclined to curse rather than to welcome the challenges with which they are confronted by the requirements of conservation—challenges which often lead to improvements and economies that might not otherwise have been won. Unfortunately the conservation of woodland was fully supported by governments only in time of war emergencies, when it was greatly outweighed by their desperate raids on the standing crop. Thus Britain was hampered from developing a strong national tradition of enlightened silviculture, and was led into a series of spasmodic efforts and narrow policies which are still a severe liability to conservation, and will take long to remedy. While nationally agriculture was little better looked after, the landowners and farmers, with less extended commitments to protect, were better able to survive their vicissitudes and to preserve the practice of conservation of the soil, of water and of shelter, which has never in Britain required large-scale governmental intervention or support. Maintenance of the fertility of the soil is indeed a normal requirement of the leases under which large numbers of farmers hold their land, and now has statutory backing.

MANY OF THE PROBLEMS ON WHICH PUBLIC CONSERVATION policies are mainly founded in other countries thus exist in Britain only in a cryptic and elusive form. This explains why it has been left until after the Second World War for an official national conservation agency to emerge, and why its approach is less in terms of the great renewable economic resources than

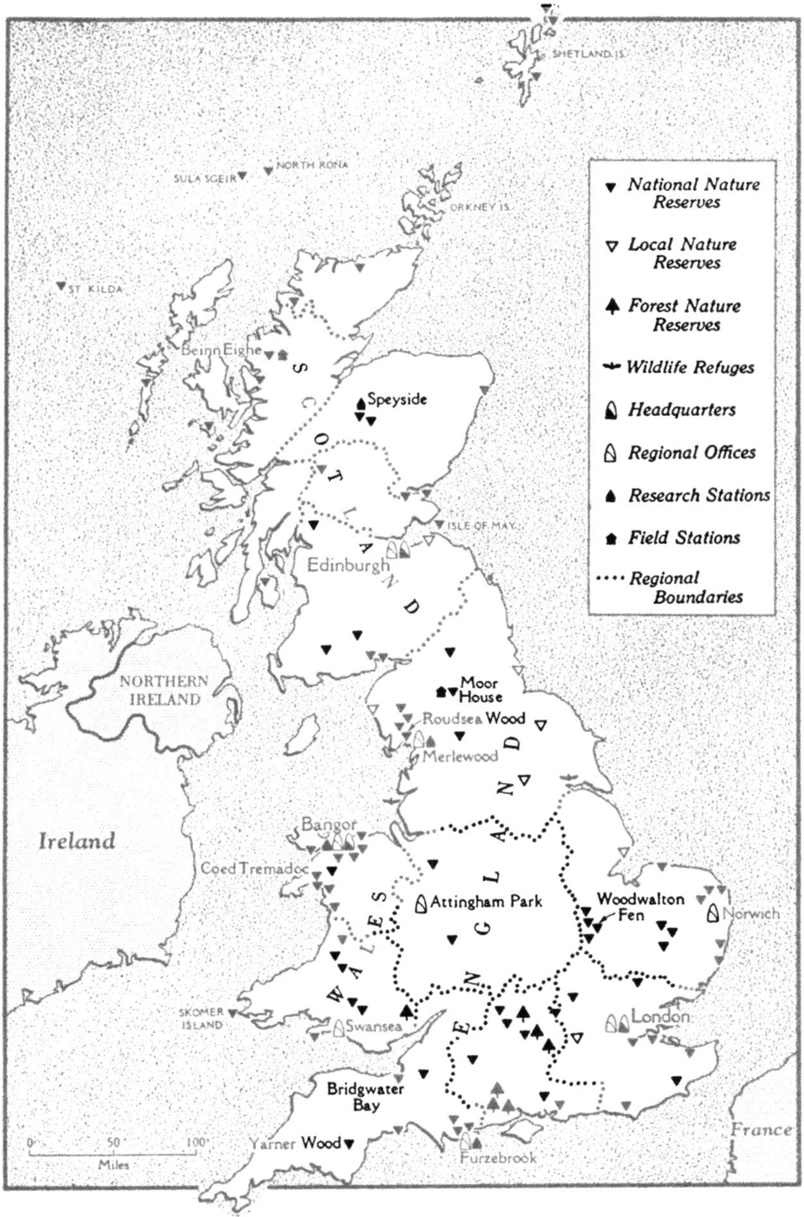

THE NATURE CONSERVANCY: Reserves, Stations, and Offices

of wild nature and wild lands. Such inspiration and impetus as there has hitherto been has derived chiefly from fears of the ruin of wild scenery, the extinction of species, cruelty to animals and the depletion of sporting stocks. Pioneer voluntary efforts such as the Royal Society for the Protection of Birds (1889), the National Trust (1895), the Society for the Promotion of Nature Reserves (1912) and the county Naturalists Trusts (since 1926) have been preoccupied largely with saving threatened sites as sanctuaries or reserves, obtaining suitable legislation and support, and educating informed opinion. During the Second World War, when numerous studies of the country's postwar needs were officially encouraged, the voluntary movement (which had previously tended to consist of a large vocal recreational and amenity wing urging National Parks, and a smaller somewhat specialized group of ecologists and others stressing the need for Nature Reserves) drifted apart. The ecologists and naturalists cast in their lot with a project, sponsored by the Royal Society, for a Biological Service to fill a gap in the provision for government support of science through the official Research Councils. By a hasty piece of surgery in 1946-49 the suggested Biological Service was grafted on as the permanent staff of the suggested Nature Conservation Board, under the convenient if in some ways misleading title of the Nature Conservancy—a name which was at once taken up also by an unofficial body of quite different and unofficial status across the Atlantic.

In March 1949 a Royal Charter brought the Nature Conservancy into existence as a public corporation and an Order in Council placed it under the supervision of a Privy Council Committee of Ministers under the Lord President of the Council, which has not yet met. Later the same year an Act, mainly dealing with National Parks, conferred on the Conservancy important powers of acquisition (if necessary by compulsory purchase) of land for Nature Reserves, which then received Crown Land status to defend it against expropriation, even if it was held not as freehold but under lease or by a statutory

Nature Reserve Agreement from some other owner. By-law making powers were also included, and the Conservancy's duties included the notification to Local Planning Authorities of any other sites, in addition to Nature Reserves, which deserved to be safeguarded on account of their special scientific interest. Under another Act these Authorities were already preparing Development Plans for their areas, on which such sites were automatically shown as a kind of use zoning, in addition to Green Belts, National Parks and certain other categories. Thus the whole of the Conservancy's territorial interests, not only in some 140 proposed National Nature Reserves but in some 1700 Sites of Special Scientific Interest were given such protection against development (a term which does not cover agricultural reclamation or tree-planting) as the British system or regional and local planning can afford. Subsequent Acts have required a number of public authorities—for instance the Central Electricity Generating Board—to have regard "to the desirability of preserving natural beauty, of conserving flora, fauna and geological or physiographical features of special interest . . . ." and have given the Conservancy powers of licensing exceptions under the Protection of Birds Act, 1954, and of representation on the Red Deer Commission set up under the Deer (Scotland) Act for the conservation and control of deer.

For what legal powers are worth, therefore, the Conservancy are armed with plenty, and most of these have proved admirably drawn and free from practical defects. Yet statutory powers, while they can lend tone, improve morale, and facilitate business, are not brandished about or used as blunt instruments by a small and junior member of the formidable governmental family entering late into a field in which the goodwill of countrymen in particular has already been seriously strained by abuses of governmental powers on the part of other agencies. The higher wisdom has dictated that the Conservancy's powers should be used sparingly, and some of them not at all. Occasionally the temptation to depart from such restraint is strong, but as time passes the forebearance of the Conservancy is giving visibly growing results in the appreciation and goodwill of many who count for much.

WHO ARE THE CONSERVANCY? WHERE DO THEY COME FROM, WHAT do they know, and how do they function? A roll call now would be answered by some three hundred men and women. Of these 18 are members of the corporate governing body. One third of these are professors, three of botany, one of zoology, one of agriculture and one of geography. Another eminent geographer, a leading soil scientist, the head of a fishery laboratory, and an ecologist who is a working vice-president of the Conservation Foundation in New York give the academic element a majority. The chairman, a prominent Scottish landowner, farmer, and forester, is also a leading amateur ornithologist and entomologist. Of the rest, two are landowners and Lords Lieutenant of counties and the other four are members of Parliament, one from each side of the House of Commons, and two from the Lords, one being the Chairman of the National Parks Commission, who serves *ex officio* for coordination, while the other is president of the Council for Nature. This recently created body is the unofficial opposite number of the Conservancy, being a confederation of over 260 national and local societies representative of the vast majority of British naturalists who belong to any organization. Much of the Conservancy's business is delegated to a Scientific Policy Committee and to territorial committees covering England, Scotland, and Wales, and these bring in 32 additional members making 50 in all, entirely unpaid, who supervise and guide the policy and program of the Conservancy. It would be difficult to name any national body or any important subject interest in the Conservancy's range which is not represented in practice, although not by nomination, among this body which through a flexible system of retirement by rotation is renewed completely within each decade, thus combining continuity with a steady influx of new blood and a wide dissemination of experience of the Conservancy's inner counsels.

The committees meet formally at about quarterly intervals, but many consultations, field visits, and working groups occur at other times. All programs and policy decisions are brought before the appropriate committees for determination.

The staff of 250 includes about 100 scientists who are engaged in scientific research and conservation. The importance attached to science by the Conservancy as a scientific body is strikingly demonstrated by this high ratio of 4 scientists to 6 nonscientists. It should be borne in mind that this residue has to allow for not only administrative, executive, clerical, and typing staff at London and Edinburgh but for land agents, foresters, and other permanently employed field staff, including wardens. In addition the Conservancy find the whole cost of maintenance of some 40 postgraduates working for doctorates in ecology and related subjects, and about as many more workers on scientific grants and contracts to universities and independent institutions. Research accounts for more than half the Conservancy's total expenditure—now just under £500,000 annually ($1,400,000 on the exchange, but the money goes farther than this would indicate—for instance the freehold purchase price of land for nature reserves has averaged little over $5 per acre).

Most of the staff scientists are botanists or zoologists with an ecological bent, but there are also biometricians, physiographers, pedologists, microbiologists, chemists, climatologists, and a geologist. The majority are based on one of the Conservancy's three main research stations or on the Scottish or Welsh headquarters; they are divided into two branches, one concerned exclusively with research, often of a fundamental character, and the other with its application in conservation and with research bearing on that.

The research program includes certain items which have been requested from outside quarters, but with these few exceptions it is designed as a comprehensive strategic attack on a range of ecological problems which are regarded as of fundamental significance for the advance of conservation. Clearly the correct choice of these problems and of the right people, sites, contexts, and resources for their successful pursuit is among the most critical of the Conservancy's responsibilities. Progress is regularly reviewed both in the field and in committee, and no reason has so far been found to make any major modification or reversal. The number of fresh projects which are brought forward backed by cogent and pressing arguments is however a constant source of worry on account of the risk of undue dispersal of effort and of becoming unable to put in sufficient extra resources to follow up successes. The Conservancy never embark on researches which are being or can be undertaken by universities and other bodies, and they hope and endeavor by their grant-aiding program progressively to encourage more outside studies in their general field. Meanwhile some preference is given to projects by which fundamental advances in knowledge can be pursued concurrently with the solution of problems of immediate practical significance for conservation, and to projects which involve and set examples of interdisciplinary teamwork, over which many institutions are handicapped by inhibitions or structural obstacles.

Which problems are in fact being given priority in research? The largest single investment (occupying virtually the whole attention of the Conservancy's largest research station at Merlewood in the Lake District of Northwest England) is in studying the biological and physical processes in which the soils and the living organisms of a woodland are involved, and comparing these with moorland which was formerly under tree cover. The point of attack is on the amounts of nutrients for plant growth which are available in the soil, the rate at which they are taken up, and the processes by which the supply is maintained. The cycle of movement of nutrient elements through the woodland ecosystem includes a regular "income" through nitrogen fixation and the addition of nutrients from weathering of soil minerals and from rain and dust. On the other hand there are losses through leaching down in the soil and removal of field crops or timber. This cycle is animated by a vast complex of processes involving the interactions of plants, animals, bacteria, and fungi, and must include alternative pathways and sub-cycles. It has already been possible to draw up balance sheets of the surprising part played by rainfall in making good supplies of such nutrients as sodium, potassium, calcium, and magnesium taken up in the tree crop. Annual dry-matter increment under broadleaved trees has averaged between 2 and 4½ metric tons per hectare (about 2½ acres), and under conifers between 3 and 18½ tons.

It is unfortunately impossible to explain such a complex program adequately in an essay of this length, but three points should be noted. Its objects are fundamental and strategic, not piecemeal or applied; it postulates a scientific controlling body able to command the confidence of mixed teams of botanists, zoologists, pedologists, chemists, and microbiologists, and the resources to employ them over long periods without interference or interruption; and it is essentially an ecological field-based effort with strong laboratory support, rather than a laboratory program involving ancillary field studies. The capacity to do such things is very near to the heart of the Nature Conservancy, and may well be found to distinguish it from all other organizations responsible for the conservation of natural areas.

Among other important studies which can barely be mentioned here are those in the dynamics of animal populations and their relation to the carrying capacity of their habitat, which are being pursued in parallel for an important game bird (the Red Grouse), an important game mammal (the Red Deer), a pest rodent (the Field Vole), and among insects for some interacting species of Ant, partly vegetarian and partly exploiting animal foods. These important population studies are a counterpart to those at Merlewood tackling the problem in terms of processes, and they are supplemented by such grant-aided investigations as those of the Wildfowl Trust on geese and ducks. Also linking up, in terms of grazing pressure, are several studies of grasslands and moorlands which are probing into the physical limits of tolerance of wild plants and the factors underlying their competition and coexistence in plant communities, the reality and nature of which is itself being analyzed. Nonbiological research includes the investigation of movements of beach material in the sea by such means as radioactive marking of pebbles, and the field measurement of evapotranspiration in relation to rainfall and runoff in different conditions of land use and land management.

In pursuing these arduous researches and in allocating to them so many resources which could readily be applied elsewhere the Conservancy are fulfilling their mission as one of the four national Research Councils mainly responsible for the adequacy and backing of civil research in Great Britain. The others, all much larger and senior, are the Department of Scientific and Industrial Research and the Medical and Agricultural Research Councils. The Conservancy regard their nature reserves as primarily living museums where scientific treasures can be permanently conserved in a flourishing state and as outdoor laboratories where they can be studied for the advancement of knowledge.

In British conditions the Conservancy's initial task is a bit like that of a director of a newly-formed art gallery whose collection is being formed largely from valuable paintings which have been left moldering in damp cellars, salvaged out of fires, slashed by lunatics, painted over in ignorant efforts to improve them, or cut up into several different pieces. Such an administrator, however aesthetically dedicated and reluctant to interfere, would inevitably find himself wanting to know more about the techniques of conservation and rehabilitation. Thus the Conservancy's scientific and conservation functions both combine to dictate a strong accent on research.

Here again is a feature in which the Conservancy as an organization is possibly unique. It is at least unusual for the same body to be engaged in fundamental research, applied research, the application of research results extensively in land management, the training of future research workers, the dissemination of information to specialist groups and to the public, and participation in the relevant policy-forming and administrative processes of government in order to ensure that the knowledge won and the dangers brought to light are taken so far as possible into account in national and local public affairs. It goes without saying that such an assignment places the whole team under constant and often acute strain, and leads to a conspicuous patchiness in its performance. On the other hand the resulting sense of urgency, of participation and of inescapable responsibility for the success or failure of conservation does something to the human beings involved which has a value beyond calculation. From the inside looking out it is difficult to measure such matters. The particular conflicts or frustrations of the moment tend to loom large, however comforting it may be to recall the Arab proverb that the dogs bark but the caravan moves on. In relation to the untidy intangibles of national public affairs even the complex ecosystem of a mixed woodland looks reassuringly measurable and manageable. Yet glancing across, as we continually must, from the science of ecology to the art of government it is not easy to see how a similar input of energy through any alternative type of mechanism might have produced bigger or better results in so brief a period as twelve years, starting with such vast arrears of knowledge, education, and management.

The work of the Conservancy is above all a scientific activity. We don't exclude the people coming to natural areas to enjoy themselves for recreation, and we recognize that the natural areas have a value as such. But the cardinal point for us is that these natural areas are treasures of science comparable to museums and comparable also to the equipment of science in other fields. Therefore, we are trying to get deeply into the scientific aspects of conservation. The organization has been set up, and its programs are carried out, on that basis. All that we do in acquisition, in imposing restrictions, and in public relations, reflects this attitude. While other aspects of natural areas are not forgotten, they are secondary.

What are the main achievements of the Conservancy and the main problems and obstacles which have emerged in pursuing them?

First, although often overlooked, is the building up of such an organization from scratch, not only without any pre-existing pattern to work upon, and with small and painfully extracted resources, but having in the process to retrain or even give primary training to nearly the entire staff from top to bottom in skills of which they had little or no previous knowledge. To recruit a newly-fledged Doctor of Philosophy and send him out alone to report on innumerable animals and plants he had never studied, against a timetable he has never experienced, with a view possibly to arguing at a site meeting with builders, farmers or engineers who have never in their lives heard of what he is talking about is not just a routine operation. Administrators having to accustom themselves to scientists being not on tap but on top, land agents having to turn their normal estate practices upside down in the interest of some form of insect or plant life, deer-stalkers having to learn meteorological recording, and foresters having to care for woodlands with the aim of never getting any produce out of them, all need not only to learn new tricks but to reorient their outlook on life. Equally at the top level the conversion of an arbitrary selection of botanists, zoologists, landowners, geographers, members of Parliament, administrators and others into a harmonious and close-knit corporation sounds easier on paper than it is in practice. Patience, human understanding, leadership, imagination, a sense of humor and many other intangibles are essential, and the Nature Conservancy has been endowed in these ways by its first and second chairmen, the late Sir Arthur Tansley and Mr. Arthur Duncan, with a magnificent human capital, to which the other unpaid members have greatly added. It is impossible to explain the organization without appreciating that it has acquired to some extent the atmosphere and friendliness of a club, whose members are linked often by more than merely official ties. For some time it was uncertain whether the Conservancy would, in view of its official status, assume the characteristics of an enlightened bureaucracy or, in view of its amateur control and its natural history affiliations, inherit the dedicated spirit of its initiators. Not without a certain struggle both these traditions have been merged, although the process is not yet complete, and the task of marrying an amateur spirit with a professional performance is one that no student of Professor Parkinson will underrate.

Second, should be placed the development of the research program. This has been discussed earlier and need not be elaborated further here.

Third, there is the conservation effort, and especially the acquisition and management of land. Most people think of the Conservancy as mainly if not entirely engaged in safeguarding nature through nature reserves. The small fraction of the public who come in direct contact with the Conservancy do so largely through the nature reserves and their wardens. Success in acquiring and protecting reserves is a measure of the effectiveness and power of the organization; success in their scientific management and use is a yardstick of the practical value of the research program; and success in winning public appreciation and backing for their acquisition and for the restrictions they necessitate is the crux of the Conservancy's public relations.

The Conservancy inherited from the previous official committees which had recommended its creation lists of natural areas which (allowing for subsequent reclassifications) totaled about 140, and covered roughly 350,000 acres. Up to the present time 85 National Nature Reserves have been wholly or partly acquired and declared with Crown Land status, most but not all having been on the original list. These total just over 140,000 acres. A further 10, totaling nearly 50,000 acres, are in the last formal stages before declaration, and there are nearly 50 others at earlier stages, most of these being relatively small. In addition, local authorities have, in consultation with the Conservancy, created 7 statutory nature reserves which are managed along similar lines, and the Forestry Commission and other public bodies have placed at the disposal of the Conservancy 7 Forest Nature Reserves. Voluntary bodies such as the Royal Society for the Protection of Birds and the county Naturalists' Trusts also operate reserves privately, often in consultation with the Conservancy, so it is hard to give a precise figure for the number of natural areas in Great Britain which are effectively being managed as Nature Reserves. On a fairly generous interpretation around 130 might now be listed, of which it will be noted, the Conservancy are responsible for managing about three-quarters, against a program of roughly 200.

IS THIS PICTURE SATISFACTORY OR DEPRESSING? BOTH VIEWS MIGHT well be argued. Ten years ago the number and extent of reserves was insignificant, the resources and quality of management were very poor and the prospects discouraging. By immense exertions a good part of the arrears has been made up, but while many irreplaceable habitats were saved in the nick of time, for others the rescuers came too late. Even now time is not on our side and the rate of fresh acquisitions has slowed from 16 in the best year, 1954-55, to only 4 last year —an indication partly of increased difficulty in negotiating the later sites, and partly of increased preoccupation with management and other problems. The view that sites not saved within the 'sixties will probably be forever lost looks uncomfortably near the truth as pressures on land spread to the last of the formerly remote or well-guarded regions, especially of the coast and inland waters.

One of the toughest problems in taking over these picked areas representative of Britain's natural and seminatural habitats has been to find out what animals and plants they contain, what changes their waters, soils, vegetation, and animal populations are undergoing, what problems have been left as a legacy of past land use and how to decide and tackle urgent tasks of protection and rehabilitation without risking further damage.

The Conservancy have devised two special tools for this purpose. The first is the Reserve Record, a permanent loose-leaf register of everything ascertained and done or published about that particular reserve from species lists and accounts of visits to specialist scientific papers, and from records of fencing, fire protection, or public control to official notices and articles in newspapers. The second, closely allied, is the Management Plan, drawn up by the responsible scientific officer on the regional conservation staff in consultation with biological specialists, land agents and administration, to digest and summarize the role which this particular Reserve plays in the national system of woodlands, moorlands, fens, sand dunes and so forth; the outline of our knowledge of its soils, climate, fauna and flora, and plans for research into them; our status and problems as owners or occupiers in relation to neighbors, public control and estate work; and the timetable and manpower, money, and equipment for future operations. As the Reserve Records build up and the Management Plans are carried out and reported on, our knowledge of what we hold and of what results from many different types of intervention or nonintervention is rapidly mounting. What is more, our previously raw and inexperienced field staffs are growing into seasoned and skilled conservationists, one of whose main problems is to swap experience from place to place as quickly as it accrues. We try to avoid so far as possible the rule-of-thumb, hit-or-miss trials so often relied on in land management, and to set up more and more of our alternative treatments as scientific experiments, accurately recorded and controlled. Some hundreds of such experiments are now being run on our Nature Reserves, and it is increasingly difficult to tell where scientific research ends and normal management begins, especially since under the Conservancy's staff structure management of nature reserves is entrusted to scientific officers with research training, who are fully interchangeable with the staffs of the research stations.

Land management now, like farming in recent decades, badly needs to check and verify its cherished practices in the light of scientific experiments. Progressive landowners appreciate that the Conservancy is in some respects their particular Research Council, able to help them on problems for which scientific advice has hitherto been lacking. The administration and staff of the Conservancy fully realize that if conservation practices are to prevail over Great Britain as a whole, this must largely come about through the vast majority of owners and occupiers of land getting to understand that the type of practices with which the Conservancy are experimenting have a valuable part to play in the management of every estate, and can be dovetailed in with the other objects of management. In this way certain Nature Reserves may come to play a role as experimental and demonstration areas for landowners, farmers, sportsmen, regional planners, engineers, and landscape architects while others will serve for higher training in ecology or (to a limited extent) for school classes to supplement the educational reserves which the schools themselves must develop. Within strict limits these things can be done without prejudice to the overriding necessity, as the American statute puts it, "to conserve the scenery and the natural and historic objects and the wildlife . . . and to provide for the enjoyment of the same by such means as will leave them unimpaired for the enjoyment of future generations." For the future all depends on the emergence of better educated, more responsible citizens who will not only appreciate their nature reserves and the work connected with these, but will accept the necessary restraint in visiting them as an obligation to their children and will be ready to give some time when they are young, as indeed members of the Conservation Corps already do, to helping with their own hands in the tasks of conservation.

The Nature Conservancy view it as their task to get in touch with all who are interested in the land and in its conservation, including the educationists, and to act as a leaven in bringing about a better appreciation of these matters and as advisers and helpers, especially for pioneer or demonstration schemes. There are encouraging signs that other public bodies, such as the British Broadcasting Corporation, the Forestry Commission and the National Parks Commission, together with such voluntary organizations as the Council for Nature, the National Trusts and local trusts and the Councils for the Preservation of Rural England and Rural Wales, as well as many people of goodwill and public spirit scattered through the country are ready to co-operate in this work. The Conservancy regard themselves as just one of the visible expressions of a growing national concern for a better future for land and scenery and soils and water and wildlife, both at home and elsewhere in the world. It is their aim by all means in their power, to strengthen the team spirit and to foster growing-points and voluntary enterprises rather than to develop as a self-contained isolationist group.

WHAT THE UNITED STATES CAN LEARN

■ EDWARD H. GRAHAM

THERE ARE SOME IMPORTANT THINGS for us in the United States to learn from the British experience of preserving natural areas—things that are apparent not only to me but to others from this country who have been in Britain as the guests of the Nature Conservancy and have seen something of the outstanding accomplishments of that organization over the short decade of its existence. First I should like to comment on certain contrasts and similarities between our two situations. Then I shall note three lessons I believe to be implicit in British experience.

Aside from the more readily recognized geographic and climatic differences between the British Isles and the United States, there is a striking contrast in land-use history. Ameri-

■

EDWARD H. GRAHAM is director of the Plant Technology Division of the Soil Conservation Service, U.S. Department of Agriculture. He has been with SCS since 1937 when he started there as a biologist. He has represented the USDA at many international conferences, including the United Nations Scientific Conference on Conservation and Utilization of Resources in 1949, the Eighth International Botanical Congress in 1954, and the last five biennial general assemblies of the International Union for Conservation. He is a fellow and past president of the Soil Conservation Society of America, and the author of half a dozen books and numerous articles in the fields of plant ecology, wildlife management, soil conservation, and land use. Mr. Graham was born at New Brighton, Pennsylvania, in 1902. He received his B.S. and Ph.D. degrees from the University of Pittsburgh.

can visitors to England usually comment on the large amount of rural countryside to be seen, since we are predisposed to think of England as a "rainy little island" with a large number of people who have lived there for a very long time. In spite of its apparent stability, the British landscape has been disturbed over a much longer period than ours. In Britain late Neolithic man ringed the trees, burned them, and cultivated the land; his cattle, goat-like sheep, and hogs kept the land open. It surprises an American to learn that most of the native forests of England were gone, not only when William the Conqueror arrived, but even before the Romans had crossed the Channel. If E. M. Nicholson can liken their task to that of the art museum director whose collection "is being formed from valuable paintings which have been left moldering in damp cellars, salvaged out of fires, slashed by lunatics, painted over in ignorant efforts to improve them or cut into several pieces," the parallel in North America would be that here we have a vast collection of magnificent originals, beautifully framed and preserved in near perfect condition, the value of which we scarcely realize and which, for the most part, we tend to ignore in our preoccupation with more practical pursuits.

In view of its history and scarcity of land, it is all the more remarkable that Britain has now established some 130 nature reserves. Most of these are small, it is true, but at least a half-dozen are more than 5,000 acres in extent, and one comprises nearly 40,000 acres. Keep in mind that in the sense normally accepted here, very few of these are "natural" areas, that is, they have for the most part been exposed to long and sometimes extreme change at the hand of man. In this country we are only beginning to realize that we stand to learn as much from disturbed environments as we do from those that have been little changed by human action.

The establishment of natural areas in itself, however, is not one of the lessons we can learn from British experience. Our record in this regard is good, although we still need protected areas that are representative of certain biotic communities indigenous to the United States. National parks, national wildlife refuges, and the wild and wilderness areas of our national forests add up to more than 40,000,000 acres. State parks, refuges, and forests add to the national network, as do private lands, such as the National Audubon Society's sanctuaries and our own Nature Conservancy's preserves. Some states have recently inaugurated programs for purchasing special sites of natural interest. Americans, for one reason or another, have had an abiding interest in preserving areas of natural landscape, and recent support for national wilderness legislation is proof that this interest is as alive as ever. What is done with natural areas, however, differs on the two sides of the Atlantic.

The governing body of the British Nature Conservancy, as Mr. Nicholson points out, has a notable variety of academic, political, and naturalist members. It provides a balance of interests in the supervisory direction of the Conservancy which we normally lack in our governmental agencies. The Conservancy, while a governmental organization, is subservient to no Department but reports directly to the Privy Council, which is essentially a committee of Ministers, or Cabinet officers. The Conservancy thus co-operates with governmental agencies at the Departmental level in its own right. Nor is the Conservancy under Civil Service. Strangely enough, in other parts of the world, as at home, one finds that a governmental body that is not under Civil Service often enjoys higher morale and achieves better performance than one that is under it. I can scarcely afford to pursue this further, however, lest I be marked as one who advocates a return to the spoils system, but the point is worth noting.

The amount of authority given to the Nature Conservancy strikes one of us as unusual in an agency of its character. The power to acquire land by compulsory purchase, to defend land against expropriation even though held under lease or agreement only, the right of notice to local planning authorities of sites needing protection, and the provision in separate Acts such as that to require the Central Electricity Generating Board to have regard to "the desirability of preserving natural beauty, of conserving flora, fauna, and geological or physiographical features of special interest . . ."—all these provisions go beyond our legislation with respect to such matters, which is limited usually to the privilege of review. As the most drastic of these statutory powers are used little or not at all by the Conservancy, however, our two systems may not in practice be as different on this score as they at first seem.

I think that many Americans will be surprised to learn of the attention being given by the Nature Conservancy to the practical application of its research. The results are plowed back into the management of the reserves themselves. The Conservancy also works closely with other governmental agencies on problems of practical land management, with a view toward applying the knowledge obtained through its research or by those working under grants awarded by it, and the Conservancy works with landowners and operators on field problems in ecology. As Mr. Nicholson points out, the Conservancy never undertakes work that can be done by others. Its research is done in the field, it deals with relationships in nature, and it is conducted in interdisciplinary fashion. Note that the British do not apologize for using the word ecology in scholastic circles, as we are inclined to do. To them it has recognized scientific status.

Appreciation of the aesthetic in nature is an attribute of the British that impresses an American. I observed an example of this characteristic in 1955 when I visited Woodwalton Fen, a 514-acre reserve leased by the Nature Conservancy from the Society for the Promotion of Nature Reserves. Woodwalton Fen is in east-central England not far from the shore of the North Sea. Surrounded by a low-lying area drained a century ago to make farm land, it was acquired as a private nature reserve nearly fifty years ago but soon lost much of its value by growing up to willow, birch, alder, and hawthorn. One of the Conservancy's ecological problems is to restore the area to the condition which originally led to its choice as a nature reserve at the beginning of this century. The great difficulty and cost of doing this has strongly impressed on present-day British conservationists the perils of laissez-faire management. In one respect, however, the management of thirty-odd years ago was far from passive. An entomological enthusiast was permitted to use the reserve to preserve a tiny orange butterfly, known as the Large Copper, which was already extinct in England, and was therefore reintroduced from Holland, where it still existed. The butterflies have survived only under artificial protection from ichneumons, which the Conservancy has carried on since it took over the management in 1954. Other examples of aesthetic appreciation by the British, such as public reaction to the planting of trees in blocks because they are considered unsightly, and the painting of transmission towers to blend with the landscape, reflect a degree of sophistication in such matters that we are only beginning to express. I think it is fair to say that the balance of attention to the aesthetic, the practical, and the scientific that is to be found in the Nature Conservancy is a combination that is not to be found in any single government agency in this country.

This brings me to the first lesson I think we can learn from the British. It is that we need a sense of what they call "amenity." By the amenities Britons mean not only an appreciation of beauty and the value of recreation but also, in the dictionary sense of the word, "the quality or state of being pleasant or agreeable." I do not think that it would have been possible to establish the Nature Conservancy during the trying years after World War II unless there had been compelling resolve on the part of many Britons that some of the world be kept pleasant and agreeable. Not only the British but many Europeans talk of landscape preservation, a term we do not use. They have in mind a landscape created by man, and of which man is a part, but they are sufficiently sensitive to the world around them that they do not want it unnecessarily disturbed. Perhaps the repeated ravages of war have made them conscious of home and environment in a way we cannot know. I am not sure how we can develop this sensitivity, except consciously to talk and teach more about the beauty of nature, and of its inner satisfactions. However it is done, we shall have to do it in our own way. We are not likely to have a counterpart of the Council for the Preservation of Rural England. But we do have a Keep America Green program, we see women's clubs fighting highway billboards, and we occasionally hear someone speak of the beauty of a strip-cropped field.

The second lesson, it seems to me, is to be found in the education received by the personnel of the Nature Conservancy. Like the appreciation of the amenities, the educational system of the British cannot be separated from the society it represents. But it nevertheless influences what the Conservancy has done. The emphasis given by the Conservancy to research, inter-disciplinary teamwork, and ecological relationships, as well as its success in establishing a facile governmental institution with a pronounced economy of administration and flexibility of arrangements for establishing reserved areas—these things result in large part from the knowledge and attitudes that have been gained from the total educational experience.

In the United States the piecemeal approach to resource allocation and use, whether in education, legislation, operation, or research, can be counted a corollary of our educational system. For professional leadership in resource fields we have leaned upon technicians—men and women who have received their education almost exclusively in the specialized field in which they are employed. You have only to review the curricula of our colleges and universities to see that this is true. You will find little that is designed to produce an educated man. The British confine the term "technician" to the laboratory assistant. There are few technicians in the Nature Conservancy. Some of its specialized personnel might deign to call themselves technologists but they think of themselves as scientists, which is what they are.

This difference is more than a matter of words. The Nature Conservancy has been instrumental recently in having established at University College, London, a postgraduate Diploma Course in Conservation Studies, modeled in part after American experience. This consists of a year's work undertaken by a student after he has received his basic education—an education that draws heavily from both the sciences and the liberal arts. Before he specializes in resources, however, the student already has been encouraged to attain a sense of history, of culture, and of society, and he often possesses some knowledge of literature, language, and economics. He is likely to have, in short, a feeling for his place in the modern world.

In the United States we are probing possibilities of turning out more rounded graduates in resource fields by adding studies in the humanities that are superimposed on the years of specialized technical training. Fortunately there is growing recognition that we have had the cart before the horse too long and that we must correct our error. Institutions being what they are, it will not be easy, but our colleges and universities, as well as employers in many segments of our economy, are beginning to demand a change. What we have done has not been good enough, and we need mightily to improve the basic education of our resource technologists.

I come now to the third lesson, upon which I should like to lay special stress. It arises from the fact that the Nature Conservancy looks upon its nature reserves as outdoor museums in which to conduct ecological, field-based research with strong laboratory support. In the United States we look upon natural areas largely for the recreational use that can be made of them. We feel that our reserved areas, whether park, refuge, or wilderness, have been established first of all for the benefit, enjoyment, and use of the people. We do not stress the scientific use of such areas. I do not wish to imply that natural areas under the administration of the Nature Conservancy are not available for use by the people, for most of them are, under certain conditions. But the view developed by the leadership of the Conservancy that these areas have high value for scientific study is not conspicuous in our consideration of natural reserves. In some cases, the Nature Conservancy considers an area so important for its scientific use that public access is prohibited.

On this point we need to realign our sights. May I say that it is not only England, with whom we are closely allied in past culture and future destiny, that appreciates the scientific value of natural areas. Countries alien to our society and purpose appreciate it as well. The U.S.S.R. has a Commission for Nature Protection that was established in the Russian Academy of Sciences in 1955. It was preceded by a Russian Society for the Protection of Nature, founded in 1924. In the U.S.S.R. there are some 60 nature reserves comprising more than 6,000,000 acres scattered across the Soviet. More to my point is the fact that the Russians have some 40 research stations on 3,500,000 acres of wild land which have complete, absolute protection from public access. These stations are devoted to the study of ecological problems.

Last summer, in connection with the Assembly of the International Union for Conservation of Nature and Natural Resources, I visited Poland. On the planned excursions the Assembly saw something of Poland's seven National Parks and 150 nature reserves. I was in the Bialowieza Park—a vast forest that bridges the Polish-Russian border. Within the forest a tract of 4,000 acres is kept inviolate for scientific analysis. We could enter it only by special arrangement. More than 100 carefully designed ecological studies were under way in this reserved area. Near Warsaw I also visited one of the five field stations of the Polish Institute of Ecology, a unit of the Polish Academy of Sciences, where intensive biological studies were being undertaken both in the field and in the laboratory by a competent staff which, incidentally, was quite familiar with the work being done in this country.

Why do these countries lay such stress on the scientific study of natural areas? The Nature Conservancy, in a way, provides an answer. First of all, its people do not draw a sharp line between research and the application of knowledge. What is learned, however abstract at the moment, is considered to have an ultimate practical value. There is recognition of the fact that the more we know about what goes on in nature the better we can cope with the problems that exist on the land, whether that land is devoted to farming, ranching, forestry, urban development, or the management of natural areas themselves. The study of natural reserves can serve as check areas, as bench marks, as controls for the practice of land use on areas of comparable soils, climate, vegetation and physiography.

It is my opinion that during the past quarter century we have sadly neglected the study of field biology. The start made a half century ago by American botanists and zoologists who studied nature, not books or machines, has given place to biometric, statistical, and molecular studies in biology which, while highly important in themselves, have been permitted to set in imbalance our concern with natural conditions and natural processes. In gaining an understanding of natural principles and concepts we have been most negligent. Such understanding cannot be attained, of course, without the knowledge that comes from detailed and widespread studies of life histories, behavior, and the relationships of living things to each other and to the environmental factors that support them.

I must add that we are, of course, conducting commendable research in field biology and that results are being published. But it is scattered, mostly segmented, and lacking large-scale encouragement, stimulation, and support. Here is the lesson of the Nature Conservancy. National leadership is needed to do the job. To provide such leadership I should like, therefore, to suggest the establishment of a Center of Advanced Study in Field Biology. The purpose of such a Center would be to provide broad leadership in, and to encourage and sponsor, by whatever means possible, the study of living organisms in nature, especially their relationship to soil, climate, physiography, and other living organisms. For such study the Center would not depend upon land and water areas under its administration or ownership, but would encourage the use of established reserves and natural areas. In addition, areas under use, as grazing, lumbering, and farming, would be utilized when appropriate. It is my belief that such a Center, in itself, should be nongovernmental in nature, although quasi-governmental status may be acceptable. Grants for research sponsored by the Center, however, might come from private or governmental sources. The Center could well be directed by a governing body comparable in status and composition to that of the Nature Conservancy.

However this proposal might be implemented, it is my conviction that leadership of this order is sorely needed. It is needed if we are to provide adequate scientific basis for solving many of the practical problems of resource use and management. It is needed if we are to maintain a competitive position with other countries that are giving much more attention to this sphere of knowledge than we are in the United States. And it is needed if we are to obtain important breakthroughs in our understanding of dynamic processes in nature, without which much of our conservation effort is uncertain and sketchy. All of these are vital objectives which the United States must resolve to achieve. It is my hope that to this, even more than to the other lessons we can learn from the Nature Conservancy of Great Britain, we might give immediate and adequate attention. To do less is not to meet the challenge that is upon us.

IV WEST GERMANY:

WATER POLLUTION ABATEMENT

ABEL WOLMAN

The Nature of the U. S. Problem

GORDON M. FAIR

● *Pollution Abatement in the Ruhr District*

EDWARD J. CLEARY

What the United States Can Learn

Archives, Emschergenossenschaft

1. ALTE EMSCHER BROOK AT HORST. This regulated stream, like other tributaries of the Emscher and the Emscher itself, carries both domestic and industrial waste waters as well as the runoff from storm rainfalls.

2. EMSCHER RIVER AT DINSLAKEN. Straightened and given a trapezoidal cross-section, this stream drains the most heavily urbanized and industrialized area of the Ruhr District.

Archives, Emschergenossenschaft

Archives, Emschergenossenschaft

3. CLARIFICATION OF THE EMSCHER RIVER AT BOTTROP. The entire dry-weather flow of the river is diverted from the main stream channel through the four settling basins that are shown in the foreground. Floating trusses support dredging pumps that lift the settled river sludge to low-lying areas that are to be filled.

4. LAKE BALDENEY ON THE RUHR RIVER NEAR ESSEN. The low-head steel-roller dam at the foot of the lake impounds about 2.4 billion gallons of water for power generation, navigation, secondary purification of waste waters, and recreation.

Archives, Ruhrverband

Archives, Ruhrtalsperrenverein

5. MÖHNE DAM ON THE RUHR RIVER AT SOEST. A high-head masonry impoundage of 35.5 billion gallons, this upland reservoir is used for water supply, power generation, and river regulation.

6. WASTE-WATER WORKS AT WUPPERTAL-BUCHHOFEN. This is one of the treatment plants for the industrial city of Wuppertal. A new high-rate activated-sludge addition is shown in the foreground before being placed in operation.

Archives, Wupperverband

Archives, Emschergenossenschaft

7. PHENOL REMOVAL PLANT. Plants such as this remove and recover most of the phenol formerly discharged into the Emscher River and thence to the Rhine.

THE NATURE OF THE U. S. PROBLEM

■ ABEL WOLMAN

MAN AND HIS WORKS have created a pollution problem since time immemorial. The more gregarious he became, the more the problem pressed upon him. Until perhaps the last century, at least in the United States, the situation was not sufficiently acute to call forth concerted attack.

Since 1860, however, volumes of public and private discussion on the mounting issues have issued from the printing presses. An impressive library is now available on almost all the facets of the subject. One might well argue that enough is known about the problem to call a halt on diagnostic national assessments and proceed to serious aspects of therapy. In corrective effort some say our country has lagged. Whatever the quanti-

■

ABEL WOLMAN is professor of sanitary engineering at The Johns Hopkins School of Engineering and the same university's School of Hygiene and Public Health, and formerly was chief engineer of the Maryland Department of Health. He is chairman of the permanent committee of sanitary engineers, Pan American Sanitary Bureau; and chairman also of the Sanitary Engineering Committee of the National Research Council's Division of Medical Sciences. He is consultant to the Atomic Energy Commission, the Detroit Metropolitan Area, the Miami Conservancy District, the Association of American Railroads, and the Bethlehem Steel Co. Mr. Wolman is a former president of both the American Public Health Association and the American Water Works Association. He was born in Baltimore, Maryland, in 1892 and received his degree in engineering from Johns Hopkins in 1915.

tative truth of this charge, the facts disclose that we have a major task ahead.

What are its characteristics? What deficiencies in approach have we crystallized? How can these be better resolved? And what light may foreign practice and experience shed upon our domestic problems?

Between 1890 and 1920 the population in incorporated municipalities grew from 25 million to 64 million. By 1925, it was estimated by Fuller and McClintock that over four times as much pollution matter was reaching American waterways as 30 years before. In 1926, these authors anticipated the familiar phrases of 1961, namely, "these increments in pollution have far outstripped reductions due to purification."

By 1935, the necessary expenditure for treatment was calculated by the Committee on Water Pollution of the National Resources Committee to aggregate a total for domestic sewage of $1.14 billion and for industrial wastes of $1 billion, both stretched over a period of 10 to 20 years.

For 1954, the Public Health Service estimated that residual sewage loads reaching streams from municipal and industrial sources had an oxygen requirement equivalent to the load that would have been created by discharge of the untreated sewage of 150 million people. The anticipated requirements for 1980 and 2000 are estimated by the same agency as 174 and 168 million; the decline in loads from 1980 to 2000 is due to assumed increases in treatment facilities. To meet these loads, some have estimated that $1 billion annually will be required.

In these estimates, the population projections employed by Resources for the Future were used at the medium rate of growth, namely, 244 million by 1980 and 329 million by the year 2000.

No matter whose estimates for cost of treatment one might favor, general agreement would certainly be reached that required annual capital expenditures will be high. Much money for this purpose has been spent in the past 25 years and much more will be spent in the next quarter century. To many observers, the pace of expenditures must be stepped up, if the country is to remove the backlog of the past 20 years and keep ahead of the urbanization and industrialization needs of the future decades.

Successful pressure for abatement at the turn of the century stemmed primarily from the emphasis on protecting water resources against disease. To the practitioner this argument is still singularly impressive, although water-borne enteric disease is rare. That the viruses offer a new but similar threat remains to be demonstrated.

The potential hazards of long-term repetitive ingestion of small amounts of exotic chemicals likewise are still to be evaluated.

Major interest in our society now turns to the protection of recreation, wildlife, and aesthetic values of the streams, lakes and estuaries. Here the man in the street, uncluttered by scientific doubts and delayed research, clamors for visible cleanliness rather than invisible risk.

This shift in public perspective may provide the energetic drive to speed up correction. It remains to be seen how effective this enthusiasm is when dollar cost confronts the citizen.

In what may be an oversimplification, it has been stated that pollution abatement has always lagged because (a) abatement works rarely benefit the offender and (b) they rarely show tangible dollar benefits. Inherent in these explanations the specific elements causing delay may be listed as follows:

Public indifference, which often may be converted into public support by militant educational programs.

Money, the competition for which in other public demands often relegates pollution abatement to secondary status.

Law, sometimes considered inadequate for orderly implementation of policy. Existence of law has not always been synonymous with successful enforcement.

Technology, rarely the cause of delay, except in special instances of industrial wastes. Improvement in technology, however, would undoubtedly spur on pollution abatement by reducing costs or increasing efficiency or both.

Political structure, or its absence. This undoubtedly delays all forms of intercommunity action, as in metropolitan areas, intercounty and interstate jurisdictions.

On the other hand, there is a persistent vein of open-mindedness in American approaches to water problems that must be taken account of in understanding the past record and assessing future prospects. Many political scientists—some, I think, invidiously—have described the United States as a country of ad hoc improvisation in its handling not only of stream pollution abatement but of water resources in general. This, to my mind, is a virtue rather than a vice because of the sheer size and complexity and variety of our problems. If we were to compile an inventory of various approaches in the general field, as well as in the particular field, throughout our country, we would be quite astonished by the variety of approaches which ad hoc responses have created from the Pacific Coast to the Atlantic Coast. It is interesting to note, for example, that the Tennessee Valley Authority, which is intended to cover almost everything, has not been the pattern since the 'thirties for any other unit in the United States. This is due again to the fact that our country is experimental. It cleaves a highway between the perfectionist and opportunist approach. All of these efforts do, it is true, exemplify a perfectly ghastly kind of world for the planner, because it is a world peopled by human beings who are always, of course, a terrific difficulty for all of us engaged in any planning activity.

THIS PREAMBLE HAS SEEMED NECESSARY TO ESTABLISH THE UNITED States setting by which to test the usefulness of foreign solutions, such as they are, as applied to the United States scene. Out of American practice pertinent questions arise across the whole spectrum of pollution abatement. These questions are posed briefly herein. Specific answers should be possible of extraction from the West German experience presented by Professor Fair in the next essay.

1. Are conditions of exploitation of water resources reasonably analogous in the two countries with respect to industrial economy, culture and urbanization?

2. Are the criteria for stream quality similar in general and specific objectives? As a corollary, what priorities of use are assigned to the streams under discussion?

3. What is the present status of stream quality and use as compared with familiar rivers in the United States, such as the Ohio, the Tennessee, the Columbia, and the Potomac?

4. What are the comparative magnitudes of the streams in watershed area, flow and variability?

5. How do the political structures and machinery created to meet pollution problems differ from those widely developed in the United States, such as sanitary districts, authorities, and TVA?

6. Are the methods of finance generally used applicable to the United States? How do their systems of taxation for the functions under discussion depart from those generally in use here? Are differences real or only apparent?

7. How much use is made of voluntary co-operative compliance effort as successfully exemplified in the Ohio River Basin?

8. What significant technologic advances, either in research or in application, are apparent in the past 25 years? Have they been paralleled in the United States?

9. Are decisions on capital expenditures made by popular referendum or by central authority order?

10. What specific lessons, in summary, emerge from the general appraisal of foreign practice? If it differs from American practice, what is the potential or the wisdom of its adaptation to the United States?

POLLUTION ABATEMENT IN THE RUHR DISTRICT

■ GORDON M. FAIR

WHOEVER TRAVELS "the wide and winding Rhine" from Cologne toward the Dutch frontier will discover that these mighty waters—averaging 500 feet in width and no less than 8 feet in depth at low water—have lost the charm of their middle distance. The vineyards on the steeps of the Rhine Massif, the chapelled villages nestling on the riverbank, and "the castled crags" of Reichenstein, and Lahneck, of Fürstenberg and Drachenfels, have long been left behind. The sightseeing steamers are gone, but commercial river traffic remains heavy. Self-propelled craft and tows of barges pursue their downstream journey for yet many miles. Then, one by one,

■

GORDON M. FAIR is Gordon McKay professor of sanitary engineering and Abbott and James Lawrence professor of engineering at Harvard University. He is Master of Dunster House at Harvard and for a number of years was dean of the faculty of engineering. He has served as consultant on sanitary engineering matters to the National Research Council and numerous other United States Government agencies, the Rockefeller Foundation, and the World Health Organization. On numerous visits to Germany, he has observed the operations of the West German river associations in a period when they were actively developing new techniques of waste-water disposal and river management. Since 1940, he has collaborated with Dr. Karl Imhoff, former chief engineer of the Ruhr River Association, in a book on the technical aspects of pollution control. Mr. Fair, who was born in the Union of South Africa in 1894, began his teaching career at Harvard in 1918. In 1951 he received the honorary degree of Dr. Ing., from Technische Hochschule, Stuttgart.

the tributaries from Westphalian lands, in particular, and the canals that pierce the high banks of the stream siphon off the heavy ore boats making their way to the blast furnaces of the Ruhr; and the empty, high-riding barges returning from Switzerland, Alsace, and Lorraine to be reladen at pit head and coke oven cast off their tow lines to slip, under their own helm, into the river ports cut into the shores of the swift-moving stream. Unashamed, the giant river has removed a mask of medieval romance to serve as the main artery of transportation to one of the most densely populated and most heavily industrialized regions of the world: the Ruhr District of West Germany, generally referred to as the Ruhr.

Seams of coal of many kinds, some with excellent coking properties, dip beneath the Ruhr on both sides of the Rhine and determine its industrial unity and potential. Coal mines, coke ovens, blast furnaces, steel mills, manufactories of heavy chemicals, textile plants, paper mills, and metal-working shops in wide variety are the principal industrial components. In spite of dense urbanization and industrialization, however, these are—except for the central core of industrial conurbations along the Emscher River—pleasant lands of rolling hills, broad fields, and mixed stands of trees. Here and there, the terrain becomes rugged enough to create favorable sites for impounding reservoirs. Some of them offer magnificent views of road and river, field and city, copse and mountain. Often there is a picture-windowed or terraced restaurant to refresh the hiker and the motorist both in body and in spirit. In truth, "Water is the eye of a landscape."

The Ruhr District comprises six river basins under the control of a unique type of water authority: the Genossenschaft. (See map, page 158, and Table 1, page 168.) From south to north on the right bank of the Rhine, the basins are those of the Wupper, Ruhr (itself), Emscher, and Lippe. More or less parallel to the left bank lies first a district (Linksniederrhein) of small streams tributary to the Rhine, and next the River Niers, which spills across the Dutch border into the Meuse.

The Ruhr District as a whole, with an area of 4,200 square miles, is only slightly smaller than the State of Connecticut, but it houses nearly three times as many people. The population density rises above 8,000 per square mile in the Emscher District and falls below 700 (and then only slightly) in but one drainage basin (Linksniederrhein). The average population density in 1960 of the whole Ruhr area was about 1,700 persons per square mile, as compared with about 800 in Rhode Island and New Jersey. The largest community in the patchwork of organized catchments is Essen, with nearly three-quarters of a million inhabitants.

To understand the institutional aspects of the West German river authorities, we must become acquainted with the structure of the West German government.[1] (For quick comprehension, this is outlined in Table 2, p. 169.) There was much the same organization at levels below that of the central government in the Kingdom of Prussia before 1918, at a time when half of the river boards were formed, and also in the Weimar Republic, when the remaining half were brought into existence. Regional political units were then provinces within states and, on the whole, somewhat smaller than the present Länder. Today the six organized river basins comprise portions of three of the four Administrative Districts (Regierungsbezirke) of Northrhine-Westphalia, which is one of the nine Länder or states, exclusive of West Berlin, that make up the present German Federal Republic. At the time of the Empire and the Weimar Republic, the Ruhr District was incorporated in part in the Rhineland Province, in part in the Province of Westphalia, both of which were major administrative areas of Prussia. Attention is drawn to this because it added to the complexity of the organizational problem of some basin authorities.

[1] For a discussion of state and local government in the West German Republic, see G. M. Carter, J. C. Ranney, and J. H. Merz, *Major Foreign Powers, the Governments of Great Britain, France, the Soviet Union, and Germany* (New York: Harcourt Brace & Co., revised ed., 1952), Ch. 7, pp. 708-16.

The visitor to the industrial complex of the Ruhr will be struck today by the absence of unsightly or neglected land and swamps, the neatness of the contained water courses, the linearity of many of the smaller rivers and brooks and some of the larger streams (more especially in the Emscher District), and the languid elegance of the wider rivers that, in the case of the Ruhr River, for example, are regulated by impounding reservoirs near their headwaters and in their lower reaches are elaborated into flights of gigantic steps of navigable pools or lakes.

This was not always so. The condition of the Emscher River, for example, which drains the central and only really grim portion of the Ruhr District, gave rise to complaints as far back as the 1600's. Then, and until the turn of the present century, this meandering stream, small but flashy, was feared for its floods. Coal mines were sunk into the Emscher Valley in the 1860's. Here and there, the ground caved in above their workings. The resulting surface depressions had no outlets. Water collected in them. Marshes formed. The flotsam and jetsam of household and industry rotted in their shallows. At the same time, the waste waters of the growing communities and industries polluted the river and its tributaries. Within the understanding of the times, outbreaks of malaria, typhoid, paratyphoid, dysentery, and other fevers were ascribed to the foulness of the river and to the black and oozing depressions that it could no longer empty.

Attempts at physical amelioration of these unhappy conditions were sporadic and of small account. There were lawsuits between lower and upper riparian abutters, and the height of absurdity in litigation seemed to have been reached in 1897 when, on complaint of downstream Altenessen, the City of Essen was enjoined by the courts from discharging its sewage into its only possible outlet channel, the Berne, a tributary of the Emscher. Told of this the chief magistrate of Essen is reported to have said that what now he would be forced to do was wall up the main sewer of Essen and let the City drown in its own effluvia. The injunction granted by the court of competent jurisdiction, along with mounting abuses of the streams, started the search for a way out of the dilemma in which most of the remainder of the Emscher District, not only its principal city, found itself. Responsible citizens and concerned administrators concluded that past solutions of the problem, both planned and executed, were quite inadequate, and that public authorities must proceed in common with mines and industry to devise and implement a new and effective master plan for the drainage basin as a whole. A meeting in 1899 was followed by the appointment of a fact-finding and planning commission which included experts from outside the region. A comprehensive plan was drawn up, adopted, and executed. Organizational, legal, and engineering studies proceeded apace. Existing water law was judged to offer inadequate relief. Therefore, a special act creating the Emschergenossenschaft or Emscher Association under public law was drafted, steered through the various legislative councils, and ultimately signed into law in 1904 by "William, by the grace of God, King of Prussia, etc., with assent of both chambers of the diet of his monarchy."

Literally, the term "Genossenschaft" means "fellowship" and implies a close association for common benefit. (For this reason I shall use the word "Association" as the English equivalent of the term "Genossenschaft.") Although the term "Verband" is used instead to denote the composition of four of the six river authorities, their legal constitution is substantially the same. This is true also for the Ruhrtalsperrenverein, in which the "Verein" component eventually implied a like organization.

The creation of the Emschergenossenschaft gave substance to the teachings of a school of jurists and political scientists headed by Otto von Gierke and committed to the principle that "society does not exhaust itself in the State, but appears at the same time in a variety of other communities, each with its own life purpose: in the family, in the church, in the commune, *in the association,* in the international community."

[Italics mine.][2] In Rupert Emerson's discussion of and quotations from Gierke's works and his general evaluation of the evolution of political and juristical thinking in the last third of the 19th century,[3] we read that the foundation-stone of Gierke's theory of the Genossenschaft was "his belief in the organic reality of human associations as persons." The corporative body to him was "no less an entity and no less a person than the individual."[4] The nature of the Genossenschaft, Gierke held to be that of a network of what he "termed 'social law,' in contrast to the public law which orders the inner processes of the State alone. The Genossenschaft is thus not only subject to law in its external activities, like the individual, but is also controlled by law in its inner life."[5]

"The actual constitutional provisions of different associations may, of course, differ very widely, but the usual organs will be a governing body or board of directors, and a general assembly of the members. The latter, Gierke is careful to point out, is by no means to be regarded as identical with the association itself. 'The individuals,' he states, 'appear in it not because of an individual right, but because they have been called together, and they transact business not as the bearers of independent individual wills seeking a contractual agreement, but as co-bearers of a common will seeking the construction and expression of a corporative decision.' "[6, 7]

These concepts were given full expression in the organization and mandates of the six river-basin authorities of the Ruhr, and it may be assumed that the interest of administrators in the writings of the contemporary legal philosophers speeded

[2] Rupert Emerson, *State and Sovereignty in Modern Germany* (New Haven: Yale University Press, 1928), Ch. 4, "The School of the Genossenschaft," pp. 126-54.

[3] Otto von Gierke, *Deutsches Privatrecht,* 1895, Vol. 1, p. 27, as quoted by Emerson, p. 129.

[4] Emerson, *op. cit.,* p. 131.

[5] Emerson, *op. cit.,* p. 133.

[6] Emerson, *op. cit.,* p. 135.

[7] Otto von Gierke, *Das deutsche Genossenschaftsrecht,* 1873, Vol. 2, p. 883.

the formation of the Emscher basin authority as a Genossenschaft, in Gierke's meaning.

The six Associations were created between 1904 and 1930 by individual "special laws" (Sondergesetze). So much acumen, however, had the Emscher Commission of 1899 brought to its task of fashioning a legally constituted and powerful administrative instrument for effective solution of the mounting problems of pollution abatement and water management in general that the special law establishing the Emschergenossenschaft became, in a wide sense, an enabling act that was followed in substance in the legislation governing all succeeding river-basin authorities both during the monarchy and after its fall.

After defining the purpose of the Emschergenossenschaft, the special law of 1904 identified its external powers and responsibilities and its internal organization and procedures.[8] Organization and procedures were further elaborated in the constitution (Statut), which was made an integral part of the law but was subject to amendment with the consent of the legally and administratively competent (zuständig) ministry.

MUCH THAT IS STATED IN THE LAWS ESTABLISHING THE SIX WEST German river-basin authorities is repeated verbatim in their constitutions. In examining the constitutional provisions we must keep in mind that the constitution determines the legal status of the Association insofar as this is not laid down in the Special Law itself. Kept in mind, too, should be the fact that the legal structure of the Association was designed to let it investigate, plan, design, construct, operate, maintain, repair, and replace all necessary installations or engineering works for

[8] The special law and statute establishing the Emschergenossenschaft are reproduced in full in the magnificent Jubilee Volume, *25 Jahre Emschergenossenschaft,* edited by the Chairman of the Board of Directors, Dr. Helbing, pp. 18-32. The meeting on December 14, 1899, of representatives of government, mining and industry to examine the question of correcting the abuse of the receiving waters in the Emscher District was declared the birthday of the Emschergenossenschaft.

the abatement of pollution in a given river basin and for the general management of its waters. It is important to note also that the Association was to accomplish its task in co-operation and co-determination with all public and private corporations or persons that were themselves polluters or drew benefits from proposed improvements; and that necessary financing was to take the form of (1) public loans for capital improvements and (2) internal allocation of running expenses to the members of the Association insofar as expenses were not covered by income.

Administrative Organs. The administrative organs of the Association are three in number: (1) the Assembly (Versammlung), (2) the Board of Directors (Vorstand), and (3) the Board of Appeals (Berufungskommission). The Associations are not special-purpose groupings of governmental units like water, fire, drainage, or metropolitan districts in which the constituent political entities and their legal and financial powers are represented as such. Nor are they fact-finding and reporting agencies like the British rivers boards, that may ask the courts for relief of nuisances but have no part in design, construction, and operation of required engineering projects. Instead, there is in the West German organizations insistence on wide autonomy, direct representation of all interested parties, whether public, quasi-public, or private, and assessment and financing not by taxation, but by direct contribution in proportion to the distribution of benefits from proposed improvements, on the one hand, and costs of preventing or rectifying damages, on the other.

The Emscher law recognizes two categories of membership: (1) that of associates or fellows (Genossen) and (2) that of participants (Beteiligte). Associates are the municipal and rural administrative districts that empty all or part of their waters into the Emscher or its tributaries. Named as participants are: (1) mines, (2) other industrial enterprises, railroads, and the like, and (3) public administrative bodies other than the municipal and rural districts, principally communes. In the Emscher Association, therefore, there is some weighting of the power of public over private corporations although this is not so great as might appear at first glance.

When the Ruhrtalsperrenverein and Ruhrverband were established as Genossenschaften, the two classes of membership of the Emscher were fused into one: that of Associates (Genossen) only. In the Ruhr Water Works Union, those proprietors of waterworks and hydroelectric plants became Associates who (1) diverted a stated annual volume of water from the Ruhr River or its tributaries, or (2) generated power, in conjunction with high-head upland impoundages, for example. Associates of the Rhurverband fall into three categories; (1) mines and other industrial enterprises for so long as their contribution to the Association equals or exceeds a stated minimum; (2) public administrative components or communes; and (3) the Ruhrtalsperrenverein. The Water Works Union is represented because its members pay for the benefits they derive from pollution abatement. Their contribution, in fact, equals 45 per cent of the annual internal budget at the present time.

The Linksrheinische Entwässerungsgenossenschaft—literally the Leftrhenish Drainage Association, and referred to commonly, and understandably, as LINEG—and the Lippeverband include the first two categories of Associates of the Ruhrverband. Special Associates of the Lippeverband are the Federal Republic, which is the proprietor of the barge canals, the State, which is responsible for the upkeep of riverbanks, and the water works and levee associations.

The Assembly is the governing organ in all the Associations. Delegates to the Emscher Assembly are elected by the common council or other elective legislative body of the member districts. Representation is in proportion to annual contribution: one delegate for each one per cent of the annual budget, but no fewer than one. A first delegate must represent the district at large; a second the mining profession; more than this number, one or the other of the categories mentioned before, each in proportion to its financial stake in the operations of the Association. The Emscher Constitution requires further that the interests of agriculture be safeguarded. By thus specifying the qualifications of the representatives, private as well as public enterprises are given a voice in the Emscher Assembly. Each category, furthermore, is heard, in some measure, in proportion to its share in the cost of operations.

In the assemblies other than the Emscher Assembly, all Associates who contribute a fixed minimum or more to the operations of the Association have a voice in the management of its affairs in proportion to their assessment, smaller contributors being allowed to pool their vote. However, no single membership category may outvote the others.

President of the Assembly is the chairman of the Board of Directors, which will be discussed later. Perhaps the most important power of the Assembly is its right to amend the Constitution of the Association, although this right can be exercised only with permission of the supervisory ministry. In addition, the Assembly reserves to itself the privilege of determining its agenda, electing the members of its Board of Directors and their alternates, approving the budgets, authorizing loans (excepting the Lippeverband), receiving the report of the Board of Directors, appointing public auditors, choosing those experts and other members of the Board of Appeals that are not appointed by the supervisory public authority, passing upon recommended changes in and additions to the general plan for basin development (Emschergenossenschaft only), establishing principles of assessment, management, and operation, formulating the conditions of employment of technical and career-administrative officials of the Association (Emschergenossenschaft and Lippeverband), and voting on proposed extensions to the area administered by the Association (Lippeverband only). Meetings of the Assembly are normally biennial.[9]

In addition to a chairman, vice-chairman, and members who, along with their alternates, are chosen by the Assembly, the

[9] Max Prüss, "Organisation und Bewährung der Wasserwirtschaftlichen Verbände im Ruhrgebiet," *Kommunalwirtschaft*, Heft 17/18, September, 1954.

chief engineer of the Association is *ex officio* a member of the Board of Directors. A medical health consultant and other experts may be engaged by the Board and invited to attend its meetings in advisory capacity. The Board, like the Assembly, must be representative of the mines and other industries as well as of the communes and agriculture. Since 1922, two employees of the participating agencies must be included.

The Board conducts all the business of the Association that is not specifically reserved to the Assembly, draws up all proposals for the meetings of that body, implements the resolutions passed by it, negotiates and justifies the budget of the Association, and renders account of all its transactions. The Board is the lawful representative of the Association, its legal competence being certified to by the supervisory public authority.

Career officials of the Association, including its chief engineer, are appointed by the Board. They are directly responsible to the Chairman who, as the principal officer of the Association, represents it in its external relationships, and manages its internal affairs. Special commissions may be appointed to study and execute individual projects under the supervision of the Board, but the technical and professional administrative staff ordinarily plan, design, construct, and operate required works.

Of wide impact is the preparation by the Board of a roster of members of the Association together with the contributions required of them. Important in this connection is the application of the principle that both the cost of pollution abatement and the value of direct, as well as indirect, benefits derived by a member from the execution, maintenance, and operation of specific installations of the Association are assessable. In the Emschergenossenschaft, enterprises other than mining are obligated in the roster only if their contributions equal or exceed 0.5 per cent of the annual budget. If they do not, the member district in which a given enterprise is situated is assessed. Association assessment then gives way to local taxation.

Assessments have the same legal status as tax liabilities; accordingly they can be collected by force, although this is rarely necessary. Nevertheless, the rules governing contributions do strengthen the rating of the Association loans or revenue bonds in the financial market. Any member deeming himself aggrieved by his assessment can ask to be heard by the Board of Directors in the first instance. Protests against its decisions can be carried, in the second instance, to a Board of Appeals on which members of the Board of Directors may not sit.

Of the nine members of the Board of Appeals in the Emscher Association, three are appointed by the public authorities, six are elected by the Assembly. The supervisory public authority names the public official who chairs the hearings of the Board. He may not be domiciled, own property, or be interested in any industry within the zone of operations of the Association. In the Emschergenossenschaft, a member of the regional Bureau of Mines, and a public reclamation engineer are the other two appointive members of the Board. Of the elected members, two are required to come from the communes, two must be associated with the mining profession, one must be engaged professionally in agriculture, and one must be an employee of an Association member. Alternates are chosen in the same way. Procedural regulations are laid down by the supervisory ministry.

As originally drafted, the constitution of the six Associations specified that there was no recourse to the courts from the decisions of the Board of Appeals. All decisions were final. Within the past decade, however, the administrative tribunals of competent jurisdiction (Verwaltungsgerichte) have refused to recognize the Board of Appeals of the Ruhrverband as a special or *ad hoc* administrative tribunal. Proper procedure, one such court has held, should leave the door open for an appeal to the courts after all other means of adjustment have been exhausted.[10] The courts will also hear cases relating to Association membership and compulsory collection of assessments. In Germany, suits among and against public authori-

[10] Prüss, *op. cit.*

ties are heard in "a separate system of administrative tribunals which, though established within the general (interior) administration, possess guarantees of judicial independence and, by and large, has given fair protection against executive arbitrariness."[11]

Public Supervision and Legislation. In company with their technical and administrative officials, regional public administrators are invited to attend, in an advisory capacity, the meetings of the Assembly and its Board of Directors. This privilege is extended also to the regional bureaus of mines.

Public supervision of the Associations is exercised in the person of a District President (Oberpräsident) designated by the administratively competent ministry. Of primary concern to him is the compliance of the Associations with the provisions of their Constitution. The minister himself, however, adjudicates complaints about constitutional procedure. Applications for loans that increase long-term debt must receive prior approval of the public authorities, and the roster of contributions cannot be published until it has been reviewed by them. General regulations governing the use and maintenance of Association installations fall into the same category. If an Association fails to post legal debts and anticipated expenditures in its budget, it may be ordered to do so, but the order may be appealed to the High Administrative Tribunal (Oberverwaltungsgericht).

The projects by which Association purposes are to be accomplished are subject to review by the supervisory ministry. For the Ruhr, Emscher, and Lippe Associations, sole supervision is currently exercised by the Ministry of Food, Agriculture, and Forestry (hereafter referred to as the Ministry of Agriculture) of the State of Northrhine-Westphalia. For LINEG and the Niers and Wupper Associations, the District President (Regierungspräsident) in Düsseldorf assumes the supervisory function, but his findings are subject to review by the Minister of Agriculture for Northrhine-Westphalia.

[11] Carter, Ranney, and Merz, *op. cit.*, pp. 718-19.

Creation of a river-basin authority under special law does not protect a polluter against civil or criminal action. The authority is not liable in his stead, nor is it so conjointly with him.

Since the formation of the Federal Republic, a Ministry for Atomic Energy and Water Management (Atomkernenergie und Wasserwirtschaft) has been entrusted with the development of Federal water policy. A water management law, passed in July 1957, offers a general framework into which possibly divergent, but it is hoped uniform, State water laws are to be fitted on initiative of the States. The Federal law itself requires application for a permit or for an acquisitive right when a given source of water is to be put to a particular use. The application may be denied, however, and there is no legal claim either to a permit or a right. Permits are revocable instruments. Rights are not; they are granted under public law. In either case, conditions of water use may be laid down and supervision of water use ordered. Penalties and fines are imposed for failure to obtain the necessary permit or right or for failure to comply with established conditions of water use. Willful noncompliance or negligence may result in imprisonment as well as fines.

Persons suffering damages from pollution may sue the polluter, and in doing so are not required to prove negligence on his part. Where the available volume of water or its quality does not meet the needs of all users, the preferred user may be required to pay compensation to those deprived of their privilege. Drinking water supplies are naturally given priority over other water uses.

General regulations may require treatment of all waste waters, biological purification of municipal sewage, and exclusion of wastes dangerous to health, of oil, of flammable liquids, and of other objectionable substances including detergents in harmful quantities. Lakes and rivers are to be kept clean enough to sustain their normal water flora and fauna, and to permit bathing.

A deadline of March 1, 1959, was set for passage of necessary State legislation, and an interstate committee was appointed to draft a model law that would be generally acceptable. The Ministry of Agriculture for Northrhine-Westphalia called the river-basin associations of the Ruhr into conference, and some of their suggestions were incorporated in the draft of a new water law that, as in the other states too, will supersede all prior legislation. Interim laws were adopted by this and some other states to bridge the period of transition. Final decisions have not yet been reached by all the states. Nevertheless, the new law went into effect on March 31, 1960.

That the states are jealous of their prerogatives is evidenced by the fact that they have attacked the constitutionality of a second Federal Law which aims to clean up the federal waterways that carry interstate and international river traffic. This case is before the Federal Constitutional Tribunal.

NOW LET US LOOK BRIEFLY AT THE MANDATES OF THE SIX RIVER-Basin Associations.[12] The original and fundamental task assigned to the Emschergenossenschaft was to convert an often sluggish, sometimes flooding, overpolluted stream and its backwaters into a series of well-regulated and serviceable waterways. To this end, flow was to be maintained at levels and velocities that would move the imposed waste loads, after minimal treatment of domestic sewage and industrial wastes, to a new mouth in the Rhine without creating either nuisance at low-water stage or damage in times of flood. It should be stated, in this connection, that a secondary, but important, purpose of the five river authorities whose streams empty into the Rhine is the protection of the Rhine itself as a major source of water for municipalities and industries below the Ruhr, both in Germany and in Holland.

In accordance with the comprehensive plan that was unfolded in anticipation of the genesis of the first German water Association, the course of the Emscher was shortened by about one-third, its terminal stretch relocated (twice in fact), and the

[12] Prüss, *op. cit.*

meandering stream and many of its tributary brooks transformed into linear, lined open conduits, triangular and trapezoidal in cross-section, and clearly identified as open drains—no longer a natural river and natural brooks. The swamps and water-logged depressions of the Emscher District are now laid dry by drainage and pumping. They have been reclaimed, about one-third of the Emscher catchment being drained by some 60 pumping stations. Water-carried settleable waste materials are removed in 24 municipal and numerous industrial waste-water treatment plants. Because the Emscher is essentially an open sewer, primary treatment is generally enough. Up to 70 per cent of the suspended solids and 40 per cent of the biochemical oxygen demand (BOD) are left behind in the sedimentation tanks designed for this purpose. The fact that sewerage of the larger communities of the district is on the combined plan and that some industrial wastes continued to find their way into the streams, ultimately led the Emscher Association to construct a giant settling plant in which the entire dry-weather flow of the Emscher is clarified. A by-pass diverts flood flows from the required basins. Coke ovens are most numerous in the district of the Emscher. To protect the Rhine against phenols, which even in minute concentration impart a vile medicinal taste to water, the Emscher Association has built some 20 phenol-removal plants at the collieries. In these, both phenol and ammonia are recovered from the coke quenching waters.

The Ruhrverband, to the south of the Emscher, had its origin in a private voluntary union of the larger water users of the region. Known as the Ruhrtalsperrenverein (literally the union of impounded reservoirs), this body was formed for the development and management, in common, of upland high-head impoundages and the safeguarding of the water supplies drawn from the surface streams within the catchment and from the alluvial gravel beds of the Ruhr valley. The Union was granted full autonomy and free control of its funds in accordance with the prescription of a legislative enabling act passed by the Prussian diet in 1879. This act was, in a large way, a

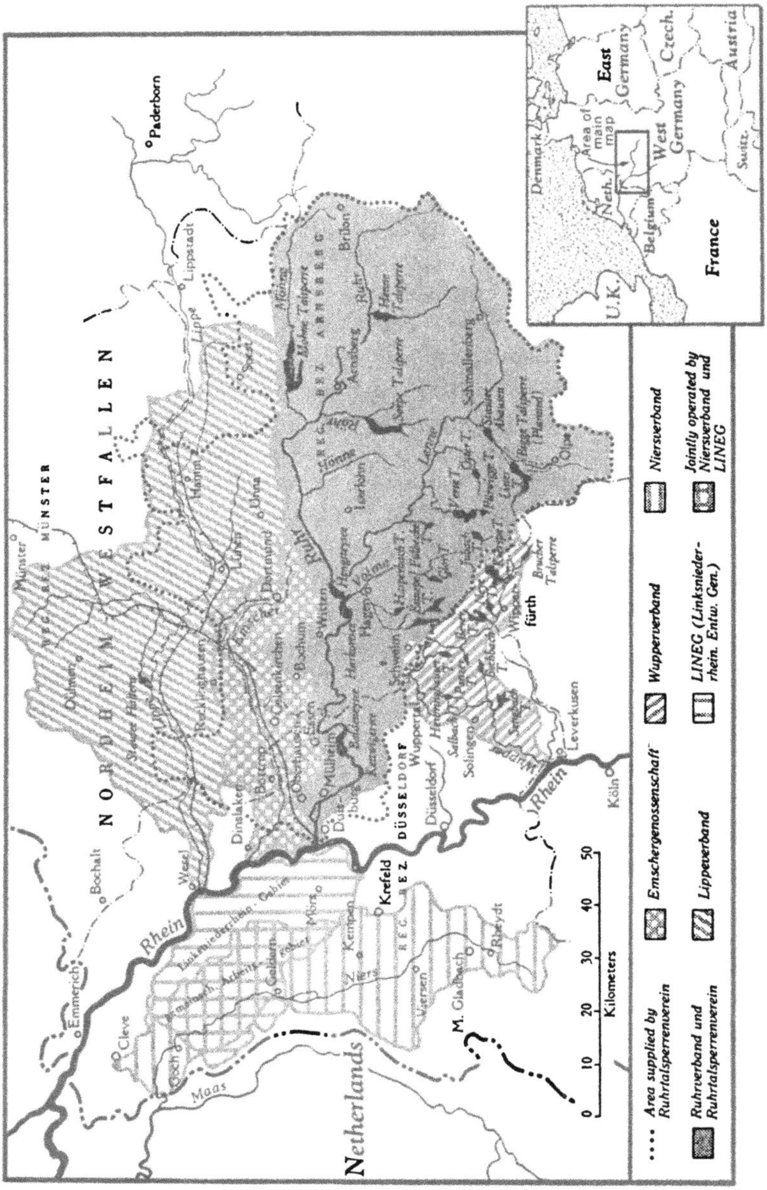

THE SIX RIVER BASINS OF THE RUHR DISTRICT

forerunner of the special legislation that created the public Genossenschaft, and the constitution of the public river-basin authorities leans heavily, and intentionally so, on the law of 1879. In 1913, the Ruhrtalsperrenverein was given public status concurrently with the creation of the Ruhrverband or Association of all polluters of the Ruhr River and its tributaries. Twenty-five years later, a merger of the two organizations placed them under single management that could give efficient expression to their complementary nature and responsibilities.

As a close associate of the Water Works Union which supplies water within the Ruhr Valley and outside, in the Emscher and Lippe regions, as well as peripheral strips along its remaining physiographic divide, the Ruhrverband has a more stringent mandate for water-quality control than the Emschergenossenschaft. To meet its objectives, the waste-water works of the Ruhrverband, in many instances, contain biological treatment units. More uniquely, perhaps, the main stem of the river has been dammed up along its course into low-head impoundages or artificial lakes. The storage provided in them, although limited, does help to iron out the hydrograph of the river. Lowering the peak flows and raising the low-water stages, the lakes, four in number, lengthen the time of passage of the river through its lower reaches and open as it were wide windows to sunlight and air. They were indeed conceived of as themselves important biological purifying agencies besides adding to the water-power and recreational resources of the region. Incidentally, the power developed in their low-head stations, including one "pumped-storage" installation, pays almost in full for the construction, operation, and maintenance of the lakes. In times of drought, some of the power is transmitted to pumping stations that lift water from the Rhine and the lower Ruhr back to its middle reaches. There it serves principally as dilution water. Some of this back-pumpage, however, may find its way into the trenches or basins from which water seeps into the gravel beds of the valley as ground-water recharge. The terminal portion of the Ruhr River is protected by a long interceptor that carries the dryweather flows from the large cities of Mühlheim and Duisburg directly to the Rhine.

At the end of 1959 the installations built and operated by the Ruhr River Basin Authorities consisted of the following:

> Ruhrtalsperrenverein—Eight upland impoundages (six more being privately owned but placed under the management of the Association) storing 75×10^9 gallons; and twelve power plants.
>
> Ruhrverband—Eighty-four treatment works, four lowland impoundages, twenty-seven pumping stations, seven back-pumping stations, six power plants, 191 miles of wastewater collectors, two pressure gas works, one phenol-removal plant, and numerous installations for the treatment of industrial waste waters.[13]

This complex of activities is perhaps the most diverse of the West German group.

The problems and objectives of LINEG, which was founded in the same year as the Ruhr Association, are more like those of the Emscher, namely, pollution abatement through main drainage. Consequently, LINEG now operates sixty-two pumping stations and fifteen waste-water treatment works.

When coal mines burrowed farther and farther north from the banks of the Ruhr River, the lower reaches of the River Lippe had to be relieved of growing pollution. Accordingly, the Lippeverband was created in 1920. For convenience and economy, especially in order to draw upon existing talent and experience, it was placed under the managemeet of the Emschergenossenschaft. The water requirements of large and busy barge canals are special problems of this region. Biological units are a part of most of the eighteen waste-water treatment works. Phenol is removed at seven installations and 46 square miles of mine-sinks are drained by seventeen pumping stations.

In 1927, a second river authority joined LINEG on the left

[13] Ruhrverband, Ruhrtalsperrenverein, Jahresbericht, 1958 and 1959.

bank of the Rhine. Known as the Niersverband, it administers together with LINEG a small portion of the middle Niers River. However, its major pollutional problems are in the upper valley where it has planned and constructed group or regional treatment works for the protection of the receiving stream. Industrial wastes have been of some concern, as has the flooding of the lower river.

Last, in time, among the basin authorities of the Ruhr is the Wupperverband which is quite like its northern neighbor, the Rhurverband, in its mission. Starting as the private Wuppertalsperrengenossenschaft, it became the Wupperverband in 1930, an Association under public law for municipal and industrial water supply and waste-water disposal and, also, for low- and high-water regulation. Pollutional problems are intensified here, as in the Niersverband by the largest municipalities and many industries being situated close to the headwaters of the central stream. A low-head impoundage is of much help. The catchment of the Dhünn River, which enters the Wupper not far from its mouth in the Rhine, was placed under the management of the Wupperverband in 1959.

Basin Hydrology and Water Use. As shown in appendix Table 3, the streams that drain the Ruhr are by no means large. The longest of them, the Lippe, traverses no more than 150 miles in a catchment area of less than 2,000 square miles. Yet the rivers of the Ruhr must provide water for town and industry, agriculture, and water transport, and must carry away the spent waters returned to them laden with wastes of household and manufactory in ever-changing composition and ever-increasing amount. How much is demanded of these streams and their guardians—the river-basin authorities—is perhaps best measured in terms of the natural flows of the different river systems. These are reported in Tables 3 to 5, in both absolute and relative terms.

The Ruhr and Wupper are mountain streams, the others flat-land rivers. The natural flows of the Ruhr and Wupper are therefore relatively large, and the opportunity for storing their flood waters in deep impounding reservoirs is relatively great.

Since mountain catchments generally contain little ground water, upland supplies of drinking water in the valleys of the Ruhr and Wupper were resorted to early in a country that normally favors ground water. Nevertheless, the prejudice against surface supplies did find expression in the development of "artificial ground water" from Ruhr River water that is led into diffusion or charging trenches and basins whence it seeps into gravel 10 to 18 feet thick that underruns much of the river valley. Wells thereupon tap these strata for their natural and augmented flows.

The annual yield of drinking water of 75×10^6 gallons per square mile from the Ruhr River is only about a quarter of the common annual yield of upland streams in the northeastern United States because the regional rainfall of the Ruhr District is about 25 per cent lower than that of the northeastern United States and, furthermore, because the reservoirs are so far upstream that only a small part of the total catchment area is tributary to the reservoirs themselves. Other purposes of upland storage in the Ruhr and Wupper valleys are flood control, power generation, and, important for pollution abatement, low-water regulation. Releases to the Wupper at low-river stage, for example, raise the ratio of river water to waste water from 0.2 to 1.0 at the river mouth.

About 17 per cent of the mean annual flow of the Ruhr River is diverted from its own catchment area to supply water to communities and industries lying within the remaining river-basin Associations on the right bank of the Rhine. Most of this water is eventually wasted to the stream channels of the respective Associations and helps to swell their flows. The Lippe loses some water to the barge canals that cross its flat lands, and the Wupper delivers some drinking and process water to the Dhünn River catchment.

The tabulation of the pollutional loads imposed upon the rivers of the Ruhr District (Table 5) will have more meaning if we recall that, to prevent nuisance, a receiving stream must carry at least some 4 cubic feet per second for each 1,000 people discharging untreated sewage into it. Expressed as a dilution factor of river water to waste water, this measure becomes numerically about 17 parts of river flow to one of sewage under present-day American conditions of municipal water use, and more nearly 35 under German conditions. In their natural state, none of the streams included in the analysis can meet these dilution requirements for untreated municipal sewage. Their flows are too small even at mean-water stage and without taking into account the population equivalent of industrial wastes. Hence low-water regulation and treatment of the waste waters have been indispensable. Taking the Ruhr River as an example, the *required* degree of purification of waste waters entering it can be calculated to approach roughly 60 per cent at mean-water flow and 95 per cent at low-water flow. On balance, the natural purification that takes place in the river stretches between points of pollution, including the considerable purifying capacity of the impoundages on the lower Ruhr River, and the dilution afforded by release of water from upland storage together with back-pumpage from as far away as the Rhine should more than offset neglect of the population equivalents of industrial discharges in our estimates.

Organization for Technical Competence. Generally speaking, Britain was the mother of rivers boards and also of the most important measures of pollution abatement by waste-water treatment because her urban and industrial growth anteceded that of other countries. The island streams were small, and the need for their conservation soon pressing. However, the basin-wide pollution-abatement agencies of the Ruhr had unique opportunities for observation and experimentation that enabled them to build upon British example, advance the art of waste-water treatment, and originate much of their own to good purpose. Especially noteworthy in municipal installations have been the separation of septic solids from the flowing sewage in Emscher tanks, called Imhoff tanks in American practice; enclosure and ventilation of trickling filters; vertical circulation of sludge masses and the collection and utilization of gas from heated sludge-digestion tanks; and uniform load-distribution combined with foam suppression in activated-sludge units.

For industrial waste waters, the extraction of phenol from coke-oven quenching liquids and the precipitation of a wide variety of waste substances from the process waters of chemical and metal industries are deserving of mention. The names of Imhoff, Bach, Sierp, Blunk, Prüss, Rohde, Schmitz-Lenders, and many other competent men are connected with this work. Most far-reaching, however, is the brilliant transformation of the Ruhr River into the four impoundments that was begun under Karl Imhoff, who was chief engineer and general manager of the Ruhrverband from its foundation and largely responsible for the fundamental simplicity of its organization.

However, one cannot know the Associations without feeling that the contributions of their engineers and scientists might not have been so significant if the conditions of their employment and practice had not been both stimulating and rewarding. Recognition of accomplishment, public esteem and honors, geographical range, diversity of projects, intensification of regional needs, are among the many factors that have made for professional and social satisfaction. No wonder entrance into the technical management of the Associations has been coveted and turnover of personnel remarkably small.

As self-governing, nonpolitical organizations, immediately responsible to their members, yet under public supervision, the Associations are free to seek the services of the most capable engineers, scientists, lawyers, and administrators.

Responsible for developing and maintaining the physical plant of each river-basin Association is its chief engineer who, as previously stated, is also a member of the Board of Directors that appoints him. Among his many duties is the negotiation of the budget and the consequent membership contributions. Ordinarily he is assisted by a deputy chief engineer who looks after basin-wide projects and matters such as general planning, stream gauging, and the collection of statistical information. District engineers supervise stream control and required structural repairs in specific portions of the drainage area. Treatment works for municipal and industrial waste waters are the responsibility of a departmental head, one of whose close advisers is the chief chemist of the Association.

A legal department is directed by an attorney, a real estate department possibly by a land surveyor, and a mechanical-electrical department by a senior man qualified in these specialities of engineering. Where coal-mine workings cause dangerous or objectionable settling of the land surface, a mining engineer is added to the staff. General nontechnical administration and accounting have a department of their own but remain responsible to the chief engineer. The supporting staff includes engineers, chemists, draftsmen, surveyors, plant operators, and clerical personnel.

The engineers are generally drawn from the pool of career men recruited to the civil service on graduation from one of the now eight (inclusive of West Berlin) technological institutions (Technische Hochschulen) of the country. After completing a rotating internship in field and office on representative public-works projects they are required to pass a special examination.

The headquarters laboratory of the Association is responsible for water-quality and treatment-plant control. Superintendents of the larger works, attendants at the smaller ones, and samplers at control points along the streams generally perform no more than simple routine tests. Sampling is held to a minimum. The gross condition of the stream, it is felt, is more revealing of what the treatment plants are accomplishing than are repetitive chemical analyses. Indeed, Karl Imhoff[14] warns that "a good operator who knows his plant and keeps his eyes open needs no meters. Plant supervision is most successful when it does not confine its attention to theoretical results of analyses and self-recorded measurements but gives expert attention to practical operations." By tradition, therefore, much of the laboratory effort is concentrated on the evaluation of new treatment processes and special studies, including theoretical and applied research.

To give an example of the number of persons engaged in the technical management of the Associations, the two Ruhr Associations had in their employ, in September 1960, 686 people, of whom about 400 were members of the technical and adminis-

[14] Karl Imhoff, *Taschenbuch der Stadtentwässerung* (Munich: R. Oldenbourg, 18th ed., 1960), p. 259.

trative staff, and the rest classified as laborers.[15] They develop and manage all of the works listed on page 160. Fifty members of the staff were university graduates, most of them engineers.

The expenditures of the river-basin authorities are sizable. Prüss[16] placed the cumulated total capital improvements in 1954 at $125 million, and Rohde[17] at $150 million in 1955. Both of them stated with pride that these funds were raised by the Associations themselves. There were no State subsidies. However, there is reference in recent annual reports of the Ruhr Association, for example, to payments by government for the repair of structures damaged in the war and subsidies from government for the provision of flood storage space in impounding reservoirs. Expenditures not matched by receipts become a legal debt of the Associations and must be met by the membership. "Ordinary expenses" include current maintenance and operation of the physical plant of the Association, interest payments, redemption of loans, and regular contributions to the replacement and contingency fund. This "self-insurance" fund covers unforeseen expenditures, more especially repair of damage caused by the elements.

Major construction is financed by loans in the form of revenue bonds. Capital improvements are considered "extraordinary expenses." Repair of physical plant damaged by mining operations is held separate as a "special expense." The total amount spent by the six river authorities during the year 1959 exceeded $54 million almost equally divided between running expenses and capital improvements. (See Table 6.)

THE SIX RIVER-BASIN ASSOCIATIONS OF THE RUHR DISTRICT HAVE now been in existence from thirty to over fifty years. During most of this time, urban and industrial growth has intensified the competition for water within their drainage areas. Pollution abatement has been of primary concern in all six. The development and safeguarding of surface sources of drinking

[15] Ruhrverband, Ruhrtalsperrenverein, Jahresbericht, 1960.

[16] Prüss, *op. cit.*

[17] Herbert Rohde, "Die Organisation und Klärtechnik der Westdeutschen Abwässerverbände," *Oesterreichische Wasserwirtschaft*, 7, Heft 1, 1955.

and process water are important responsibilities in two; more particularly so because the supplies from these two are carried into neighboring basins as well. In all these years, the work of the Associations has continued without letdown. It may prove to be timeless.

One of the river-basin authorities, the Emschergenossenschaft, has been in action, almost without constitutional change, for over half a century, under no less than four sovereignties: a constitutional but authoritarian monarchy, a strongly centralized federalist republic, a ruthless dictatorship, and a decentralized federal republic. Survival has, patently, been through strength.

The long service of the Associations betokens, above all, their usefulness to their local communities, their industries, and the State. They continue to flourish as autonomous, pluralistic organizations with representation of both private and public interests in their constituent organs. Assessments are levied in proportion to the costs of abating pollution caused by individual members of the Associations and the distribution of benefits received by members from physical improvements.

The only major constitutional change has been the validation of appeal to the courts when grievances cannot be settled by the two appellate organs within the Associations themselves, a natural development in a growing democracy.

A Special Law and a Constitution govern Association conduct. Public supervision or review guarantees compliance with legal and constitutional prescription. The elective or representative process is involved in determining the composition of every organ. Unique is the fact that all corporations affected, whether private or public, are represented either directly or indirectly in the councils of the Associations. Unique, too, is the power to plan, design, build, and operate needed installations and to do so with funds voted directly by the Assembly of the Associations and assessable against the polluters and beneficiaries of pollution abatement. The fact that few assessments are protested is evidence of the confidence placed in these Associations both by the general public and the business community. In the final analysis, however, the success of the Associations depends heavily also on the competence of their technical and administrative staff.

The genius of a people expresses itself in varying ways. The German river-basin Genossenschaft appears to be one of the best.

Acknowledgments. The author is grateful to many friends for supplying much-needed information, more particularly, to Dr. Karl Imhoff, "dean" of waste-water engineers in the Ruhr District and in the wider world; Dr. Herbert Rohde, director of the department of waste-water treatment in the Ruhrverband; and Dr. Berndt Dieterich, desk officer for stream pollution abatement in the Ministry for Atomic Energy and Water Resource Management in the Federal Republic.

APPENDIX

TABLE 1. River basin authorities of the Ruhr District[1]

Year of organization	Name	Area of jurisdiction (*sq. mile*)	Population 1956 (*thousands*)	Population density (*persons/ sq. mile*)	Responsibility
1904	Emschergenossenschaft	326	2,630	8,070	River regulation and pollution abatement
1913	Ruhrtalsperrenverein[2]	2,590	4,550	1,750	Water supply and quality management
1913	Ruhrverband[2]	1,740	2,090	1,200	River regulation and pollution abatement
1913	Linksniederrheinische Entwässerungsgenossenschaft[3]	374	260	695	River regulation and pollution abatement
1926	Lippeverband[4]	1,070	1,030	962	River regulation and pollution abatement
1927	Niersverband[3]	514	575	1,120	River regulation and pollution abatement
1930	Wupperverband[5]	239	752	3,180	Water supply and quality management, river regulation and pollution abatement
Total		4,260	7,330	1,720	

[1] Population estimates by Karl Imhoff, *Gas- und Wasserfach*, Heft 18, 1956.

[2] The two Ruhr River authorities were placed under the same administration in 1938. The Ruhrtalsperrenverein (water-supply authority) supplies numerous communities in the Emscher and Lippe districts and small areas along the southwestern and eastern boundaries of the Ruhr Basin. Thus its area of jurisdiction and its population are not counted in the totals because both are fully accounted for by the figures for the Ruhrverband and the portions of other areas supplied with water. It was organized as a private association in 1899.

[3] The two left-bank authorities jointly administer an additional area of 162 square miles.

[4] The headwaters drainage area of 814 square miles is not organized. The lower drainage area is administered by the Emschergenossenschaft. The estimated population of the total basin is 1,250,000.

[5] The Wupperverband had its origin in the privately organized Wuppertalsperrengenossenschaft (Impounding Reservoir Association) incorporated in 1896.

TABLE 2. Governmental subdivisions of the German Federal Republic

Levels of government	Units of government	Organs of government
Central or Federal (Bund)	West Germany	a. President (Bundespräsident) b. Chancellor (Bundeskanzler), and Ministers c. Diet (Bundestag) and Council (Bundesrat)
Regional or State (Land)	States (Länder) 9 exclusive of Berlin	a. State President (Ministerpräsident) b. Ministers c. Diet (Landtag)
Intermediary (In large states only)	Government or Administrative District (Regierungsbezirk)	a. District President (Regierungspräsident)
Local	1. Municipal District or City-County (Stadtkreis) 2. Rural District or County (Landkreis) 3. Commune (Gemeinde) City, Town, Village, and, in some areas, a District Board of several towns or villages (Amt)	1a. Chief Magistrate or Lord Mayor (Oberbürgermeister) 2a. District Councilor (Landrat) 3a. Magistrate or Mayor (Bürgermeister) 1, 2, and 3c. Common Council (Rat) varying in different regions

Note: The authorities and lines of local supervision are from the lowest to the highest: (a) Mayor of Town or Village - via Mayor of Amt, where it exists - to District Councilor of Rural District; (b) Councilor of Rural District and Lord Mayor of Municipal District respectively - via District President where an intermediary Administrative District exists - to the Minister of Interior of the State.

Source: After G. M. Carter, J. C. Ranney and J. H. Merz, *Major Foreign Powers, the Governments of Great Britain, France, the Soviet Union, and Germany*, New York, Harcourt Brace & Co., rev. ed., 1952.

TABLE 3. Hydrology of streams in the Ruhr District[1]

Item	Emscher[2]	Ruhr	Lippe[3]	Niers[3]	Wupper[4]
Catchment area, sq. mile....	303	1,740	1,890	560	239
Natural flows					
Low water, cfs...........	35	140	140	42	18
cfs/sq. mile............	0.12	0.08	0.07	0.07	0.08
Mean flow, cfs...........	250	2,600	1,100	210	490
cfs/sq. mile............	0.83	1.5	0.58	0.37	2.0
Mean annual flow and storage					
Mean annual flow, 10^9 gal..	58	620	260	50	120
10^6 gal/sq. mile........	190	360	140	89	500
Storage in upland reservoirs, 10^9 gal...........	—	75	2.3	—	16
10^6 gal/sq. mile........	—	43	—	—	67
Length of stream, mile....	44	146	147	73	71

Note: The Rhine river distance between Cologne and the Dutch border is 105 miles. The discharge is 36,000 cfs at ordinary low water, and the velocity at mid-stage is 2 to 4 fps. At maximum flood stage, the flow is 440,000 cfs.

[1] Flows tabulated by Helmut Möhle (Wupperverband) Wasserwirtschaftliche Probleme an Industrieflüssen, *Die Wasserwirtschaft*, 45, 4, 1954. LINEG is not included.

[2] Before the Emscher River was straightened, it was 68 miles long.

[3] Values are for the complete basin of the Lippe and Niers rivers.

[4] Excluding the Dhünn River.

TABLE 4. River-water use in the Ruhr District[1]

Item	Emscher	Ruhr	Lippe[2]	Wupper
Annual water use:				
Drinking water, 10^9 gal.............	—	130	16	12
10^6 gal/sq. mile.................		75	8.5	50
Process water, 10^9 gal..............	—	16	—	5
10^6 gal/sq. mile.................		9.2		21
Cooling water, 10^9 gal..............	15	120	172	7
10^6 gal/sq. mile.................	50	69	91	29
Total, 10^9 gal......................	15	260	190	24
10^6 gal/sq. mile.................	50	150	100	100
Percentage of mean annual flow used...	25	43	70	20
Diverted from catchment, 10^9 gal.......	1	110	16	8
10^6 gal/sq. mile.................	3.0	63	8.5	34
Percentage of mean annual flow lost....	1	17	6	7

[1] Source of information as in Table 3; the Niers River is not used as a source of water. Needed supplies are drawn from the ground.

[2] Additionally 53 x 10^9 gal. annually are diverted into canals.

TABLE 5. Pollutional loads of streams in the Ruhr District[1]

Item	Emscher[2]	Ruhr	Lippe[2]	Niers	Wupper
Annual waste water production, 10^9 gal.					
Domestic, 10^9 gal.	32	16	12	5	9
Industrial, 10^9 gal.	48	24	17	7	13
Total, 10^9 gal.	79	40	29	12	22
cfs.	340	170	120	49	95
mgd.	220	110	81	32	61
Population, 1000's	2,400	1,500	1,250	550	700
Available dilution (cfs/1000 population)					
Natural low-water flow	0.015	0.093	0.11	0.076	0.026
Natural mean-water flow	0.10	1.7	0.88	0.38	0.70
Ratio of river water to waste water					
Natural low-water flow	0.1	0.8	1.1	0.8	0.2
Natural mean-water flow	0.8	14	9.1	4.2	5.3

[1] Source of information as in Table 3.

[2] The natural flow of the Emscher and Lippe is increased by the water supplied to these drainage basins from the Ruhr River. The natural flow of the Ruhr River is reduced correspondingly.

TABLE 6. Annual budgets of the river basin authorities in the Ruhr District, 1959 (millions of U.S. dollars)[1]

Name	Ordinary budget[2]	Extraordinary budget[3]	Special budget[4]
Emschergenossenschaft	3.48	1.85	3.35
Ruhrtalsperrenverein	8.93	12.76	—
Ruhrverband	6.26	4.56	—
LINEG	1.28	1.81	—
Lippeverband	1.59	1.85	0.44
Niersverband	1.79	2.00	—
Wupperverband	1.53	1.23	—
Total	24.86	26.06	3.79
Grand Total	54.7		

[1] From a table compiled by Professor Paul Gieseke (Bonn University) for the Economic Commission for Europe.

[2] Operation, maintenance, repairs, replacements, interest, and sinking funds.

[3] Capital improvements.

[4] For repairing damages caused by mining operations.

WHAT THE UNITED STATES CAN LEARN

■ EDWARD J. CLEARY

HOW FULLY MIGHT THE EXPERIENCE in the Ruhr Valley be applied to situations prevailing in the United States? If I were obliged to sum up in one sentence what we learn from the Ruhr Valley it might be this: When people are faced with a crisis—but generally not until then—they respond with ingenuity and purpose.

This conclusion will not be comforting, particularly to those with a perfectionist turn of mind. I would describe the perfectionist in this instance as one who envisions an orderly world in which all things are planned so that contingencies can be met before they occur. The frustrations of the perfectionist arise because individuals and groups are seldom distinguished in doing what obviously is the best thing to do. This lament

■

EDWARD J. CLEARY for the past dozen years has been executive director and chief engineer of the Ohio River Valley Water Sanitation Commission, with headquarters at Cincinnati. After four years as a practicing civil and sanitary engineer, he was on the editorial staff of the Engineering News Record from 1935 to 1949, serving as executive editor from 1945. At various times he has served as consultant to the Institute of Inter-American Affairs, the World Health Organization, and the Federal Republic of Germany. He is a clinical instructor at the University of Cincinnati Graduate School of Medicine, and former president of the American Public Works Association. Mr. Cleary was born at Newark, New Jersey, in 1906. He received his B.S. from Rutgers University in 1929, his M.S. in 1933, C.E. in 1935, and Sc.D. in 1959.

finds eloquent expression in the poems of Omar Khayyam, notably in the quatrain wherein he cries:

> Ah Love! could thou and I with Fate conspire
> To grasp this sorry Scheme of Things entire,
> Would not we shatter it to bits—and then
> Re-mould it nearer to the Heart's Desire!

A half-century ago the situation in the Ruhr Valley, so far as water resources were concerned, presented a "sorry scheme of things." Expansion of industry coupled with concentration of population in a region of limited and unregulated water resources promised economic strangulation. In brief, the Ruhr region faced a crisis. Professor Fair has graphically portrayed the conditions that existed—great swamps caused by coal mine subsidence, flash floods, the flotsam and jetsam of municipal and industrial wastes rotting on the shores of streams, and grossly polluted sources of water supply.

It was this "sorry Scheme of Things" that provided the opportune time and setting for tackling the problem of water-resources management in order to "Re-mould it nearer to the Heart's Desire." And this remolding found expression in the establishment of river-basin authorities with broad and unique purposes and powers.

Thus, as pointed out by Professor Fair, in 1904 "William, by the grace of God, King of Prussia, etc., with the assent of both chambers of the diet of his monarchy" gave his blessing to the people of the Ruhr Valley to work out their salvation in the management of water resources. How well the people of the Ruhr responded to this challenge has been documented by Professor Fair. The results claim admiration of the entire world. And we might wonder why the example of these river-management boards has not found wider application in other countries.

In pondering this lack of emulation as applied to situations in the United States, at least two reasons suggest themselves.

1. Despite claims of an apparent crisis in water resources, conditions are not yet serious enough to create incentives for

aggressive action to deal with them. This view, of course, harmonizes with my basic assumption that when there is a real crisis people do find effective means of dealing with it.

2. Government actions and arrangements are predicated on prevailing social views and values. Thus, in seeking to understand why there is little application of principles that have guided the destiny of the Ruhr Valley we must acknowledge a vast change in social outlook as contrasted with that of the 1900's. For example, the concept that it is the responsibility of a local region to devise, apply and finance solutions for its problems no longer is considered realistic. This is the era of the welfare state, wherein it is rationalized by theory and made effective by principles of taxation that somebody other than the individual, or the community, must share, if not completely accept, burdens that once were considered personal or local. In many cases the burden has been shifted to the central or federal government.

An example in the United States that illustrates a form of crisis in one phase of water management, and the change of attitudes in dealing with it, lies in the realm of flood control.

Let us turn back to the morning of March 23, 1913. On that day, paralyzing disaster enveloped the Miami River valley in the State of Ohio. It marked the beginning of a flood that caused the loss of 300 lives and property damage exceeding $100 million. From this calamity arose one of the most magnificent demonstrations of courage, self-reliance, and democratic achievement in the pages of American history.

Ten years from the time when much of the valley lay under water, the flood-control works of the Miami Conservancy District were completed to prevent the recurrence of such a catastrophe. Who provided this guarantee? It was the people of the Miami valley—under local initiative and at local expense. In so doing the citizens of this area pioneered in developing new legislative implements, new practices in organization, new engineering principles; as time has shown, they made the valley safe from the devastation of floods. (Those who are interested in details will find them set forth in the book *The Miami Conservancy District,* by Arthur E. Morgan, published in 1951 by the McGraw-Hill Publishing Co.)

Since that grim day in 1913 there has been a vast change in the attitude of the American people toward flood-control responsibility and who pays for it. Regardless of local circumstances, flood control is now a well-established national burden. Today, a call for local action and local financing for protection from floods would be quite incomprehensible. Small wonder, therefore, that the experiences in the Ruhr Valley and the example of the Miami Conservancy District in the United States might be greeted with something less than enthusiasm when we discuss river-basin organization and administration today.

Another more recent manifestation of how far we have departed from the concept of local responsibility in the solution of water-resources problems is the insistence on federal subsidy for the construction of community sewage-treatment works. Under the claim of a national water-pollution crisis and the indisposition of municipalities to clean up their sewage discharges, the federal government five years ago undertook to provide grants-in-aid. The purpose was to stimulate construction of treatment facilities in "poverty-stricken" communities. It now appears that with regard to building sewage-treatment works every community is poverty-stricken; and legislation has now been introduced to more than double the federal subsidy. One is led to presume, therefore, that if a crisis does exist it is not yet apparent to those municipalities who are responsible for it.

WITH THIS BACKGROUND WE CAN POSE THE QUESTION: WHAT ARE the possibilities for applying experiences in the Ruhr Valley to situations in this country?

We might first agree that philosophically and politically we are not too well disposed to embrace the structure and principles of river-basin administration as practiced in the Ruhr.

Not even in Germany—where the Ruhr Valley demonstration is close at hand for all to study and applaud—do we find significant emulation of this successful pattern. This would seem to strengthen the argument that regardless of its virtues people are not motivated to adopt a proven course of action until a crisis compels some sort of change.

So far as the United States is concerned, the trend has been that it is proper—and of course cheaper—for a locality or region to seek solutions in federal actions. Witness, for example, the attitude recently expressed by the General Assembly of Indiana. On February 23, 1961 the legislature passed, and the governor signed, a petition calling upon "the Congress of the United States to enact effective legislation for the control of floods and the prevention of soil erosion along the rivers and streams within the State of Indiana to preserve natural resources, protect and promote the health, safety and general welfare of the people, and to appropriate sufficient funds for the execution of such legislation."

This attitude—that flood control is a logical problem, with attendant costs, to be transferred to Washington—is not peculiar to the citizens of Indiana. Many examples of similar viewpoint could be cited; the Indiana petition simply happens to be the most recent expression that is conveniently available to me. It would appear, therefore, that the prevailing mood in the United States is not at all hospitable to the reception of ideas and principles that have been evolved for water-resources administration in the Ruhr.

There is, however, one countervailing trend to federal paternalism that has important implications. It is represented by an unspectacular but steady growth in the employment of interstate compacts for the resolution of regional problems. The interstate compact movement offers promise of a resurgence in the acceptance of local responsibility. The Port of New York Authority, established by the states of New Jersey and New York to cope with the crisis of transportation strangulation in a metropolitan area, is an example. In spirit, as well as in fact, the origin, purposes and accomplishments of the

Port of New York Authority would compare favorably with the German water associations although its field of endeavor is quite different.

Much more restricted in scope and powers are the half-dozen interstate agencies engaged in regional pollution-control programs. For example, the Ohio River Valley Water Sanitation Commission, with which I am connected, resulted from a compact among eight states that have joined together solely for the purposes of supplementing and co-ordinating their activities for pollution control. It has nowhere near the scope of powers of the Ruhr associations nor does it have the same comprehensive responsibilities for water-resources development. The Ohio River commission, for instance, cannot raise money for the building of sewage-treatment plants; its role primarily is one of co-ordinating action among its signatory states, each of which retains its sovereign status in redeeming its pledge of co-operation. Establishment of such limited-purpose organizations is, however, a manifestation by the people in a region that they recognize a problem and will try to solve it without burdening the federal government. And it might be added that students of political science and water-resources management have credited these agencies with being a vital force in the co-ordination and promotion of local and regional efforts.

The most exciting prospect—and one in which we may find some parallels reminiscent of the Ruhr Valley undertakings—relates to the Delaware River basin. Early in 1961 the draft of an interstate-federal compact was submitted to the Congress of the United States and to the legislatures of Delaware, New Jersey, New York and Pennsylvania for approval and ratification.

This compact would establish a regional agency to comprehensively administer all aspects of water use and development in the Delaware Valley, as well as to promote sound practices of watershed management. Thus the people of the Valley, as represented by the governors of four states and the mayors of New York and Philadelphia, the two largest cities with vital interests in water supplies, have asserted that they are prepared and willing to cope with the crisis of managing a limited resource.

This is not the place for a detailed analysis of the Delaware Valley compact. Suffice it to say that the proposal does represent a strong affirmation of acceptance of local responsibility. And it establishes a mechanism for the determination of policies and the execution of desires by those most vitally concerned—the people of the Valley. With ingenuity it provides for the co-ordination of existing federal-agency interests in the Valley—some twenty-five in all—but does not defer to federal jurisdiction; in brief, it recognizes the federal government as a full partner but not necessarily the sole arbiter of destiny or source of all funds.

The following recital of some of the general powers of the proposed Delaware River Basin Commission may suggest to readers, as it does to me, the view that principles governing the Ruhr Valley associations have found expression, at least in part, in the Delaware River compact. Among other things, the Delaware Basin agency may:

> Plan, design, acquire, construct, reconstruct, complete, own, improve, extend, develop, operate and maintain any and all projects, facilities, properties, activities and services, determined by the commission to be necessary, convenient or useful for the purposes of this compact.
>
> Negotiate for such loans, services or other aids as may be lawfully available from public or private sources to finance or assist in effectuating any of the purposes of this compact; and to receive and accept such aid upon such terms and conditions, and subject to such provisions for repayment as may be required by federal or state law or the commission may deem necessary or desirable.
>
> From time to time, after public notice and hearing, fix, alter and revise rates, rentals, charges and tolls and classifications thereof, for the use of facilities which it may own or operate and for products and services rendered thereby, without regulation or control by any department, office, or agency of any signatory party.

And in order to promote integration of viewpoints the commission may:

> Constitute and empower advisory committees, which may be comprised of representatives of the public and of federal, state, county and municipal governments, water-resources agencies, water-using industries, water-interest groups, labor and agriculture.

So far as I know, the drafters of the Delaware River Basin compact were not indoctrinated with the experiences of the German water associations. Nevertheless, it would appear that in their independent assessment of the Delaware basin situation and the means for dealing with it they have reached conclusions quite in harmony with principles that have found favor in the Ruhr Valley.

There has been little understanding in the United States—and even less appreciation—of the Ruhr Valley programs. Not until Professor Fair was invited by Resources for the Future to address himself to the matter have we had such a profound analysis of the associations, their mode of conduct and their contribution to water-resources management. And the introductory comments of Dr. Wolman provide an appropriate setting for the presentation of the Ruhr experience. Together, the two essays make available a fund of detailed knowledge that permits an evaluation of German river-basin administration.

One must be aware of a basic difference that emerges from a comparison of the Ruhr Valley and American approaches to pollution control. The Ruhr concept has favored consideration of pollution control as an integral element of water-resources management rather than as a separate entity. In fact, German terminology expresses this inclusive relationship in the word "Wasserwirtschaft," which connotes water management or perhaps, better still, the total water economy.

Only recently in the United States have we been emphasizing that pollution-control practice bears close relationship to all the varied aspects of water-resources economy—that flood-control reservoirs may also be employed to augment dilution in low-flow periods; that hydroelectric plant operations need to be integrated with pollution-control requirements, or that navigation improvements may exercise a profound effect on water-quality conditions. Traditionally in the United States pollution control has been identified, and in large measure, administered, primarily as a public-health function. Lacking the broader view of relating pollution-control practices to water-resources management in general, our administrative, financing and technical approaches reflect marked differences from those employed in the Ruhr.

It is no simple matter, therefore, to translate Ruhr Valley experiences into American practice. Such adaptation would call for quite an overhauling of deeply-rooted philosophies and existing institutional arrangements. It is for that reason that we may regard the proposed Delaware River Basin Compact as offering an exciting prospect. And it might not be out of order to suggest that what happened in the Ruhr Valley a half century ago is finding expression in the Delaware Valley today.

V CANADA:

LOCAL INTEGRATION OF LAND AND WATER PROGRAMS

D. A. WILLIAMS

The Nature of the U. S. Problem

EDWARD G. PLEVA

• *Multiple Purpose Land and Water Districts in Ontario*

AYERS BRINSER

What the United States Can Learn

1. FANSHAWE DAM, Upper Thames River Conservation Authority. Designed primarily for flood control, the installation also provides a permanent recreational lake of 650 acres.

2. STREAM BANK PROTECTION, Grand Valley Conservation Authority. Gabion groynes on Whiteman's Creek, built by the Authority for experimental use. The groynes are made of heavy gauge rust-proof wire baskets hand filled with rocks.

3. HEART LAKE, Metropolitan Toronto and Region Conservation Authority. On some warm Sundays more than 7,000 people visit this recreation area.

4. FLOOD DIVERSION CHANNEL, Metropolitan Toronto and Region Conservation Authority. Photograph shows spring flood waters being safely discharged around low-lying central business section of the town of Brampton.

5. **FARM POND.** One of the more than 2,400 such ponds built during the past fifteen years with the technical and financial assistance of Conservation Authorities.

6. **EXTENDING A GAME BIRDS' HABITAT,** Saugeen Valley Conservation Authority. Fifty Hungarian Partridges being released at a site deemed suitable for the species, which already is well established in some other areas of Southern Ontario.

7. MASS TREE PLANTING IN THE GANARASKA FOREST. Boy Scouts turn out for the weekend under the auspices of the Ganaraska River Conservation Authority. Fourteen Authorities have agreements with the Minister of Lands and Forests for establishing and managing Authority Forests. Abandoned land unfit for agriculture is bought by the Authorities and restored to forest production.

8. CONTOURING AND STRIPCROPPING. Conservation Authorities co-operate with Soil Advisory Service of Ontario Agricultural College to interest farmers in combatting erosion and encouraging them to carry out soil conservation measures.

All photographs courtesy of the Conservation Branch, Ontario Department of Commerce and Development

THE NATURE OF THE U. S. PROBLEM

■ D. A. WILLIAMS

TO PROVIDE A SETTING for Edward Pleva's account of experience with multiple-purpose districts in Ontario, let us review briefly some of the principles and problems underlying the operation of districts—wherever they may function—in the conservation of soil and water and related natural resources.

I have a deep-seated faith in the efficacy of local organizations such as the soil conservation districts in the United States and the somewhat similar conservation authorities in southern Ontario. It has been my privilege to work with these districts in the United States as a technician or in administrative capacities since their inception nearly a quarter of a century ago,

■

DONALD A. WILLIAMS is administrator of the Soil Conservation Service, U.S. Department of Agriculture, a post he has held since 1953. He entered SCS in 1935, and from then until 1947 held field and regional SCS jobs as a conservation engineer with emphasis on water conservation and irrigation. He was assistant regional director of SCS in the Pacific Coast Region from 1947 to 1950; flood control survey officer, Office of the Secretary of Agriculture, 1950-51; and assistant Chief of the Soil Conservation Service, 1951-53. In 1960, at the invitation of the Indian Government, he conducted a survey of the soil and water conservation needs of India. He has received the Distinguished Service Award of the National Association of Soil Conservation Districts and the Distinguished Service Award of the U.S. Department of Agriculture. Mr. Williams was born at Clark, South Dakota, in 1905. He received his B.S. degree in civil engineering from South Dakota State College of Agriculture and Mechanic Arts. He later did postgraduate work at the same college and at the University of South Dakota.

in 1937; and I have followed the development and the substantial accomplishments of Ontario's conservation authorities since their beginning in 1946.

Today, in our own and other areas of the world that have been occupied and developed by rapidly growing numbers of people, land faces multiple and often competing uses—for agriculture, living space, industry, wildlife, forestry, recreational and other purposes. In order to achieve harmonious multiple use of all land resources, it is accordingly necessary to approach the problems involved on a broad, multiple-purpose basis. Emphasis must be given to decisions, planning, and action by land users themselves, on the individual farms, in conservation districts, in communities or municipalities, and in watersheds or river basins.

One of the strong features, I believe, of the Ontario approach to conservation of land, water, forest, wildlife, and mineral resources has been the breadth of authority vested in their conservation authorities. In the United States, unfortunately, there too often has been a tendency to fragment the resource job by creation of new special-purpose mechanisms for every facet of the task. We have drainage districts, grazing districts, irrigation districts, flood control districts, and many others.

It was not until our soil conservation district idea was conceived about twenty-five years ago that the broader multiple-purpose conservation and land use mechanism began to materialize. Today, our soil conservation districts, for example, are the principal sponsors of watershed protection and flood prevention projects, having recognized this as a logical extension of the farm-by-farm approach to soil and water problems. They are carrying out, within their district framework, significant activities in drainage, irrigation, land reclamation, woodland management, wildlife habitat improvement, and many other facets of resource management because these are, indeed, along with erosion control, integral parts of the total picture of soil and water conservation. Districts are now beginning to explore their role in the area of rapidly rising urban-rural problems, with a special hope that they can help to achieve, among other things, better order in the rapid conversion of land to nonagricultural uses. Even so, it is my opinion that our soil conservation districts are still not exercising the breadth of authority and responsibility desirable in the interests of maximum attainment of the conservation objective.

The main reason for the success of the soil conservation districts program in the United States is, I am sure, their emphasis upon local leadership directly exercised by farmers and ranchers elected as governing officials by landowners and operators of their own districts, which in turn have been established by the vote of district area landowners as authorized by State enabling legislation. This is a slightly different concept of local leadership from that found in Ontario where representatives of municipalities conduct the program of action of conservation authorities.

It may be helpful at this point to consider briefly the mechanisms employed elsewhere for carrying forward soil and water resource conservation programs. Such programs have been initiated in some fifty countries. Three of them—Canada, New Zealand, and the United States—are outstanding for their patterns of local approach. In each case, I credit their considerable progress to the use of some local mechanism—catchment conservation districts in New Zealand, conservation authorities in Ontario, and soil conservation districts in the United States. This noteworthy achievement is in contrast to the slower or more limited resource conservation gains made in many of those countries where the pattern has been entirely one of national or federalized management programs.

An effective approach to resource conservation problems anywhere includes at least three fundamentals: (1) the existence of a district, authority, or other local mechanism; (2) the availability of competent technical assistance; and (3) an effective information and education program to make people aware of conservation needs and keep them abreast of conservation opportunities and developments. There is, of course, the need of financial support from private and public sources. The role of these fundamentals is well illustrated by the experiences of Ontario in Canada and of New Zealand and the United States, which reveal significant parallels, as well as some differences born of problems encountered and their respective political genesis.

In all three countries the district mechanisms are local government organizations, though variously initiated and established. All have governing bodies constituted of local people, the majority of whom are either elected directly or, in the case of the Ontario conservation authorities, appointed from elected municipality representatives.

All of these districts or authorities are concerned with natural resource conservation, but with variations in original purpose, scope of their activities and subsequent points of emphasis. Thus the soil conservation districts in the United States originally gave primary emphasis to carrying on farm and ranch soil and water conservation programs, but since the mid-1950's have become increasingly active in watershed protection, flood prevention, and water management work in community-type water problems. The Ontario conservation authorities and New Zealand's catchment conservation districts, on the other hand, were conceived primarily for watershed flood and pollution control, and for river control, respectively, but subsequently have given increased attention to erosion control and other land-use work. In all three countries, basically important conservation farm planning is practiced.

All of them make their own decisions and carry out their own programs, with the advice and technical or financial assistance of federal, state, or provincial agencies. This principle of operations, significantly, is the reverse of that under which local boards or other bodies function merely in an advisory capacity in the implementation of federal or state programs.

On the other hand, there are important differences between the three types of resource conservation mechanisms. One is in staffing. The soil conservation districts and conservation authorities generally utilize the technical and other personnel of government agencies. The catchment conservation districts have their own staffs, with their plans and work subject to review by a central government staff.

Soil conservation districts for the most part have no taxing power, although a number of states have authorized creation of sub-districts with such authority, which is limited to the sub-district boundaries and is used principally for watershed project financing. New Zealand's districts have power to tax directly for both conservation works and administrative expenses, and the Ontario conservation authorities levy taxes through the municipalities, primarily for conservation works. The United States soil conservation districts receive contributions in money, services, and other forms from many sources, including state funds appropriated in varying amounts by different states.

Although they have the authority under state law to do so, soil conservation districts do not handle subsidies. The Ontario and New Zealand districts do. Although districts in two-thirds of our states have authority to make land-use regulations, such authority seldom has been used, because the concept of voluntary participation in the conservation program has been felt by virtually all concerned to be essential to achieving sound and lasting soil and water conservation, except in isolated emergency situations. The New Zealand districts likewise can make land-use regulations, but the Ontario conservation authorities do not have this authorization.

Especially worth noting, I think, is the attitude in the three countries' districts toward information. Both Canada and the United States found from the outset that it was necessary to keep people abreast of conservation developments. As a result, people in both countries regard soil and water conservation as basic to the economy and well-being of their country. While recognizing the great amount of river control work and notable advances in erosion control made in New Zealand, responsible observers feel that many of the problems encountered there could have been forestalled to a large extent had an information and education program been regarded from the beginning as essential to conservation development.

Soil conservation districts now include more than 90 per cent of the land in farms and 95 per cent of the farms and ranches in the United States. A total of 1,850,000 landowners and operators are now co-operating with the districts. The experience we have had with these instrumentalities of soil and water and allied resources conservation has been gratifying indeed.

When these districts started, the inexperienced farmer-supervisors understandably were not prepared always to assume full immediate leadership, and it was necessary for agencies, especially the Soil Conservation Service and Extension Service, to do some of the things for them that they properly should do themselves. As they grew in experience and confidence, however, the district officers, with less and less guidance, assumed the responsibilities of true self-leadership that today characterizes the majority of soil conservation district governing bodies.

The more opportunities to serve their communities that present themselves to the districts, and the more responsibilities that are directed their way, the better able they seem to be to take on the larger tasks in stride. It is important to keep in mind, also, that the problems of soil and water conservation are quite different and more complex now than they were even a few years ago. The flexibility of the district mechanism has proved itself in the way in which these districts have kept their attitudes and activities in pace with the broadening demands upon them. Expanding watershed development, making adjustments to continuing urbanization of agricultural lands, and other major problems and trends challenge the districts as never before in their key position in resource development.

The community concept of resource conservation has been growing apace in the United States. Not only conservationists but responsible leaders throughout government, business, educational, and other fields feel fortunate that we have in the soil conservation districts the machinery needed to deal with a multitude of problems stemming from use and management of land and water as resources for living. They are at the hub of the wheel, so to speak, and can reach out to pull together the available public and private facilities required to get the job done.

MULTIPLE PURPOSE LAND AND WATER DISTRICTS IN ONTARIO

■ EDWARD G. PLEVA

IN MANY CHARACTERISTICS of soils, climate, and population, Southern Ontario, the area with which we are primarily concerned here, has much in common with neighboring sections of the United States. Most of our problems, no less than our accomplishments and opportunities, will be familiar to readers south of the U.S.–Canadian border.

Southern Ontario has flood hazards and local difficulties with water supply, soil erosion, needs for farm drainage and for reforestation; rapid growth of its cities accentuates demands for outdoor recreation and creates difficulties of land use on the

■

EDWARD G. PLEVA is professor and head of the Department of Geography at the University of Western Ontario. He has been connected with the University since 1938, when he came to start courses in geography there. He is President of the Canadian Association of Geographers, Examiner-in-Chief in Geography for the Ontario Department of Education, and consultant in geography to various governmental departments and ministries. He is editor of the Canadian Oxford Atlas and of the McGraw-Hill Canadian Geography Series. He is specially interested in community planning, regional planning, and river valley development, both professionally and as a participant in organizations in his area. Mr. Pleva was born in Minneapolis, Minnesota, in 1912. He received his B.A. and Ph.D. degrees from the University of Minnesota.

rural-urban fringe. Nearly all of the land is in private ownership, much of it in farms. Relationships between government and the individual are central to a diversity of programs of financial or technical assistance to landowners.

There is little that is unique, then, in Southern Ontario's problems of land and water development or in the physical techniques of dealing with them. Much more distinctive, I believe, is the multiple purpose land and water resources district as a device for carrying out a coherent set of programs that draw upon the experience and skills of several departments of the provincial government. For those reasons I shall devote most of this essay to an account of the structure and functions—the formalities, if you will—of Ontario's multiple purpose local Conservation Authorities and an examination of the Conservation Authorities Act under which they were established and are operating. It will be useful, however, to begin with a glance at Southern Ontario as a whole, as the geographical background of these activities, and to close with a brief sketch of other legislation that bears on present and prospective land and water programs under the Conservation Authorities Act.

Ontario, the central province of Canada, lies near the economic heart of North America. Its southern border adjoins a populous and highly industrialized area of the United States.

Ontario, with a total area of 412,562 square miles (80 per cent of which is land), is Canada's second largest province. It is half again as large as Texas and nearly five times the size of Great Britain. It accounts for one-half of the country's industrial output and one-third of her population—about 6 million out of a total of 18 million people. Ontario is heavily urbanized and predominately English-speaking, although there are important pockets of French-speaking population in the northern and eastern sections. More than any other part of Canada, southern Ontario resembles the major northern and midwestern areas of the United States.

The province is made up of two dissimilar areas of unequal size—northern and southern Ontario; the natural division between the two sections occurs where the French River, Lake Nipissing, and the Mattawa River cut across a relatively narrow extent of land between Georgian Bay and the Ottawa River.

Northern Ontario has 93 per cent of the province's total area and 12 per cent of its population. It is in general a land of potential and actual mineral, forest, and recreational wealth. A rigorous climate, rocky terrain, and scarcity of agricultural soil combine to preclude agriculture throughout most of the area. The great bulk of the land is still in public ownership, consisting for the most part of Crown lands administered by the provincial government.

The southern section, with about 29,000 square miles and more than 5 million people, contains only 7 per cent of the area of Ontario but 88 per cent of the total population. It is in the main an area of level to undulating sand and clay plains and rolling hills, and it has a climate suitable for agriculture. Nearly all of the land is privately owned. At present about 80 per cent of the people of southern Ontario live in cities or towns of more than 5,000. The remainder is divided about equally between farm and rural nonfarm families. And the towns and cities still are growing fast. In another decade the problem of fringe development may well have spread over nearly all of the area. It is almost entirely in southern Ontario that work under the Conservation Authorities Act is being carried out.

THE OPERATION OF MULTIPLE-PURPOSE LAND AND WATER resources districts in Ontario is related to the Conservation Authorities Act, passed by the Ontario Legislature in the early part of 1946.

The Conservation Authorities Act permits the local municipalities within a drainage basin to organize as an Authority for purposes of flood control, reforestation, land-use programs, fish and wildlife management, farm ponds construction, wet lands programs, and any other scheme related to the conservation of natural resources within a drainage basin. In

Ontario, a municipality is, by definition, a city, town, rural township, or county. The municipality, in effect, is the unit of local government. Under this system we have found that rural interests, on whose lands many of the programs are actually carried out, are quite adequately represented.

The first step toward setting up a Conservation Authority must be taken by the municipalities wholly or partially within a watershed. Although the initiative for organizing the Authority must come from the local municipalities, the Act sets out clearly the provincial responsibility at the organizational and subsequent stages.

The process of organization is relatively simple. Two local municipalities within a given watershed must first by resolution petition the Minister of Commerce and Development to call a meeting to determine whether an Authority should be established. The Minister then sets the time and place of the organizational meeting, to which representatives from all the municipalities are invited. The representation is on a population base, as follows: municipalities below 10,000, one representative; over 10,000 but not over 50,000, two; over 50,000 but not over 100,000, three; over 100,000 but not over 250,000, four; over 250,000, five. Two-thirds of the representatives must be present to make the meeting legal.

At the organization meeting, the legislation is explained and the merits and limitations of the proposed Authority are discussed. Finally the vote is taken and if two-thirds of those present are in favor, a resolution is sent to the Minister asking that an Authority be established. The Authority is made legal by an Order-in-Council and under the Conservation Authorities Act becomes a body corporate with representatives from all the municipalities within the watershed or drainage basin. The membership includes those municipalities which voted against its establishment as well as those municipalities which may have abstained from voting or were absent at the organization meeting. In brief, membership by the municipalities in the watershed is total and complete.

The establishing of an Authority is, then, a simple legal

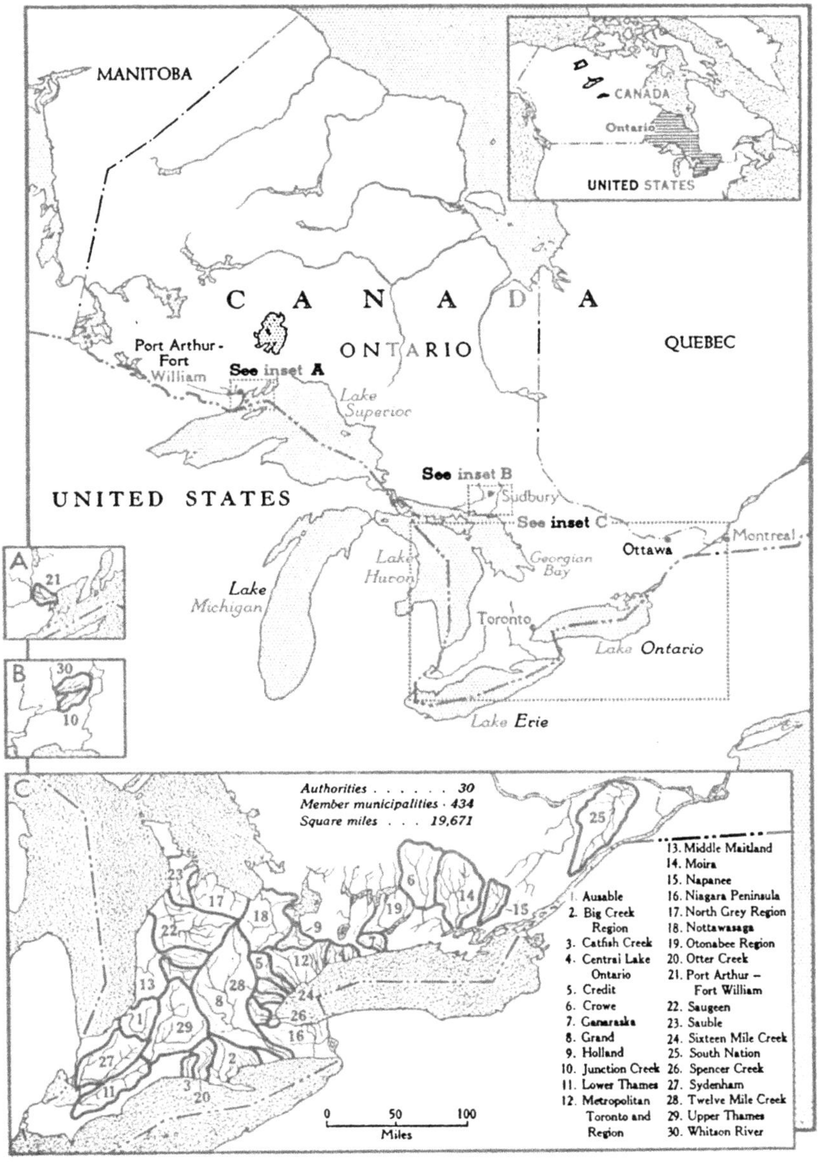

CONSERVATION AUTHORITIES IN ONTARIO

matter. Since the Act was first passed in 1946 not one request for establishing an Authority has been refused.

At the beginning of 1961 there were thirty Authorities in Ontario. Together, they covered an area of 19,671 square miles. Total membership of 695 included 434 different municipalities. The number of members from a given municipality is based on the same population formula as for representation at the organization meeting. In one district—the Grand River—there also is a Commission, which antedates the Authorities Act and continues to carry out responsibility for flood control.

The Authorities vary greatly in size from one with an area of 86 square miles and 8 members to one with an area of 2,614 square miles and 78 members.

The powers of an Authority as set forth in Section 15 of the Act are as follows:

> To study and investigate the watershed itself or by its engineers or other employees or representatives, and to determine a scheme whereby the natural resources of the watershed may be conserved, restored and developed and the waters controlled in order to prevent floods and pollution or any of such matters; subject to the Lakes and Rivers Improvement Act, to erect works and structures and create reservoirs by the construction of dams or otherwise; to purchase or acquire and without the consent of the owner enter upon, take and expropriate any land which it may require and sell or otherwise deal with such land or other property; to purchase or acquire any personal property which it may require and sell or otherwise deal therewith; to enter into such agreements for the purchase of materials, employment of labor and such other purposes as may be necessary for the due carrying out of any scheme; to enter into agreements with owners of private lands to facilitate the due carrying out of any scheme or conservation project; to determine the proportion of the total benefit afforded to all the participating municipalities which is afforded to each of them; to use lands which are owned or controlled by the authority for such purposes, not inconsistent with its objects, as it deems proper; to acquire lands, with the ap-

proval of the Minister, and to use lands acquired in connection with a scheme, for recreation purposes, and to erect, or permit to be erected, buildings, booths and facilities for such purposes and to make charges for admission thereto and the use thereof; to collaborate with departments and agencies of government, municipal councils and local boards, and other organizations; to plant and produce trees on public lands with the consent of the Minister of Lands and Forests, and on private lands with the consent of the owner, for any purpose; to cause research to be done; generally to do all such acts as are necessary for the due carrying out of any scheme.

Surveys and Reports. The early Authorities set up in Ontario were brought into existence because of the recurring threat of flooding. However, all Authorities are aware of the necessity of comprehensive and integrated measures in a regional conservation program. The Authorities themselves were not staffed or equipped to carry out the necessary studies in land use, reforestation, woodlot management, pollution control, water supply, fish and wildlife, and recreation. The Ontario Department of Commerce and Development, through the Conservation Branch, carries out at no expense to the Authority the necessary preliminary surveys and investigations. The technical staff of the Conservation Branch then appraises the conservation needs of each drainage basin or watershed. Finally a detailed report is submitted to each Authority outlining the Conservation measures that should be followed.

The report is in the form of a working plan and is intended primarily for Authority members. The full report tends to be a rather bulky document since it contains fairly detailed information on the watershed. A typical report for one of the larger watersheds runs to 600 pages, 100 maps and charts, 150 pictures, and 75 specific and detailed recommendations. In addition to the full report, a summary report in printed form is issued for general distribution throughout the watershed and is especially valuable for the schoolteachers. Every schoolroom has a copy of the local survey on its reference shelf.

The full report is based on survey work and is written up under six general headings: History, Land Use, Forestry, Water, Wildlife, and Recreation. The scope of the studies varies with the conditions and needs of the area investigated. The findings reported to the Authority have direct application to the problems the Authority must face and solve.

History: A background of history is necessary to give the proper perspective to immediate problems. The residents of the watershed must have as localized a picture of past conditions as possible. The historical approach often promotes an interest in conservation among people who would otherwise be indifferent.

Land Use: The relations between soil, forestry, agriculture, and water are carefully studied. All available information dealing with this relationship within the watershed is drawn together. Soil surveys by the Soils Department of the Ontario Agricultural College and physiography studies by the Ontario Research Foundation provide the basis for detailed field work and aerial photograph interpretation.

Forestry: The condition and extent of the original forest, the history of the wood-using industries, the conservation measures in progress, the inventory of existing forestry resources, and the recommendations for future conservation measures make up the forestry section.

Water: A careful examination is made of all hydrometric and meteorological records for the watershed. Much attention is given to flood records. A careful hydrology survey indicates the number, size, and location of reservoirs needed to control floods and to regulate summer flow. The survey includes a mapping and examination of all lakes, ponds and old mill dams.

Wildlife: General inventories are made of all species of wildlife, both game and non-game. Special emphasis is given to vanishing or threatened species. Streams are classified for fish management.

Recreation: The present and projected future populations of the area are studied and the facilities needed now and in the future are estimated and described.

A survey of a given watershed may take several years to make. The field work is done largely during the summer months when students from the universities carry out the investigations under the direction of the provincial scientists and technicians. A typical survey may give professional training and summer employment to thirty students.

The Authority, when presented with the report, must assume responsibility for undertaking the schemes recommended. The Authority usually initiates action on those projects it considers most urgent; in some cases, however, the Authority may take many years before starting on any part of the program. The legislation throughout is permissive, and even after an Authority has been set up, the carrying out of the program depends on the favorable action of the Authority.

Operation and Financing. If one or more schemes have been approved by an Authority, it becomes the Authority's responsibility to make approaches to the government departments or other bodies from which it requires assistance, whether financial, technical, or otherwise. A land-use program would involve the assistance of the Department of Agriculture; a forestry or wildlife program, the Department of Lands and Forests; a flood-control project, the services of an engineer accredited by the Department of Commerce and Development and the Department of Public Works, who does the engineering and designing up to the point of calling for tenders and carrying the work through to completion of the structures. The Authorities, that is, are not staffed for specialized conservation work. Neither is the Conservation Branch of the Ontario Department of Commerce and Development which works most closely with them. But the technical assistance of the operating departments is available to the Authorities, and they use it.

The financing provisions of the Conservation Authorities Act recognize three classes of expenditure. The first is related to capital expenses such as dams, reservoirs, reforestation land, and conservation areas such as parks and floodways. The Authority's share of payment for capital expenses is borne by the member municipalities which benefit from the scheme, on the basis of a fixed percentage of estimated benefits. A second class of cost is related to the maintenance on capital projects and is paid entirely by the Authority in the same way. A third class relates to administration costs and includes all the things an Authority might be expected to do that do not fall into the categories of capital or maintenance costs. The Authority meets its share of these costs by charging the municipalities proportional shares based on the assessment of lands within the drainage basin. Administration costs include salaries, office rent and equipment, tree-planting and other portable conservation machinery, exhibits, audio-visual materials, continuing investigations of farm ponds, reforestation lands, and other small conservation projects, printed material and traveling expenses.

Grants are made by the Ontario government for the first and third classes: capital expenses and administration costs. Grants have not been made so far for maintenance of capital projects. Grants are a matter of policy and may change from year to year. At the present time the grants from the Ontario government for each of the following is 50 per cent: a flood control scheme costing less than $5 million, a large-scale reforestation land purchase, a conservation area in which a park is located, the acquisition of flood-plain lands, and all items included under administration costs.

A flood-control scheme which costs more than $5 million is eligible for a 37½ per cent contribution of the Government of Canada under The Canada Water Conservation Assistance Act. The Province of Ontario contributes an additional 37½ per cent of the cost, leaving the Authority with a 25 per cent responsibility.

Receipts coming to the Authority from nontax or nongovernmental sources, such as camping fees or boat rentals, are used for the development and expansion of the general progress of the Authority. Receipts coming from the municipalities are raised by taxation and show clearly in the mill-rate levy on every landowner's tax bill. Each municipality collects and turns over to the Authority the amount of money represented in its levy. Grants of the senior levels of government are paid to the Authority and do not go through the treasuries of the member municipalities.

Even so brief a review as this of the cost-sharing provisions of the Act suggests four points worth noting. First, the local municipalities through the Authority must initiate the programs and are responsible for a significant share of the financing. Second, the Government of Canada is under no legal obligation to assist the province in conservation work because the terms of the British North America Act of 1867 specifically place the control of natural resources under the jurisdiction of the provinces. In practice this has meant that the benefiting area must feel convinced that the project is worthwhile enough to warrant a considerable outlay of local money, and also that the case for assistance must be presented and defended adequately to warrant the expenditure of money at the higher levels of government, particularly from the Government of Canada.

Third, although the Act has plenty of teeth in it, to be used if necessary, we have thus far in practice been able to keep things from coming to the point of coercion. For instance, the law clearly provides for expropriation of land, if necessary, in carrying out an approved program. Neither the officials nor the private citizens like the word "expropriate." A few years ago, when 5,000 acres had to be acquired for the Conestogo Dam, the Authority negotiated sales with the landowners with such success that the indemnity price of only four or five out of ninety-four parcels had to go to arbitration.

Fourth, for work that benefits private land as well as contributes to an area program—like tree-planting, farm drainage, and the building of farm ponds—the farmer himself must contribute, usually through hiring labor or supplying it himself, or through buying trees that the government offers at a nominal price.

All important decisions are made by the full Authority. Many of the large Authorities have executive committees that carry out the routine work. An important feature of the Conservation Authorities Act is the provision for the appointment of advisory boards or committees for any subject considered necessary to the effective work of the Authority. Membership on an advisory board is not limited to members of the Authority, and consequently the people with special or technical skills in the area are put to work. Subjects dealt with by such boards or committees include: flood control, conservation, education, public relations, farm ponds and little dams, reforestation, land use, water pollution, parks and recreation, and historical sites and properties. Final decisions must always be made by the accredited members of the Authority but often it is the work of the various boards and committees that brings the meaning of conservation and land use to every person living in the drainage basin.

The Upper Thames Authority. The Upper Thames River Conservation Authority is an example of what can happen in a drainage basin under the legislation. The Authority was organized by Order-in-Council on September 18, 1947, and is one of the oldest Authorities in Ontario.

The Authority was organized primarily to develop a program of flood control on the Thames River. There has been a steady evolutionary development of the conservation idea in the watershed so that flood control today is just one aspect of an integrated river valley development program. After the provincial officers completed the survey shortly after the organization of the Authority, a series of meetings and conferences led to the first major project, the construction of Fanshawe Dam upstream from the city of London on the north branch of the Thames River. Fanshawe Dam was built with contributions from three levels of government: Canada 37½ per cent, Ontario 37½ per cent, and the local municipalities

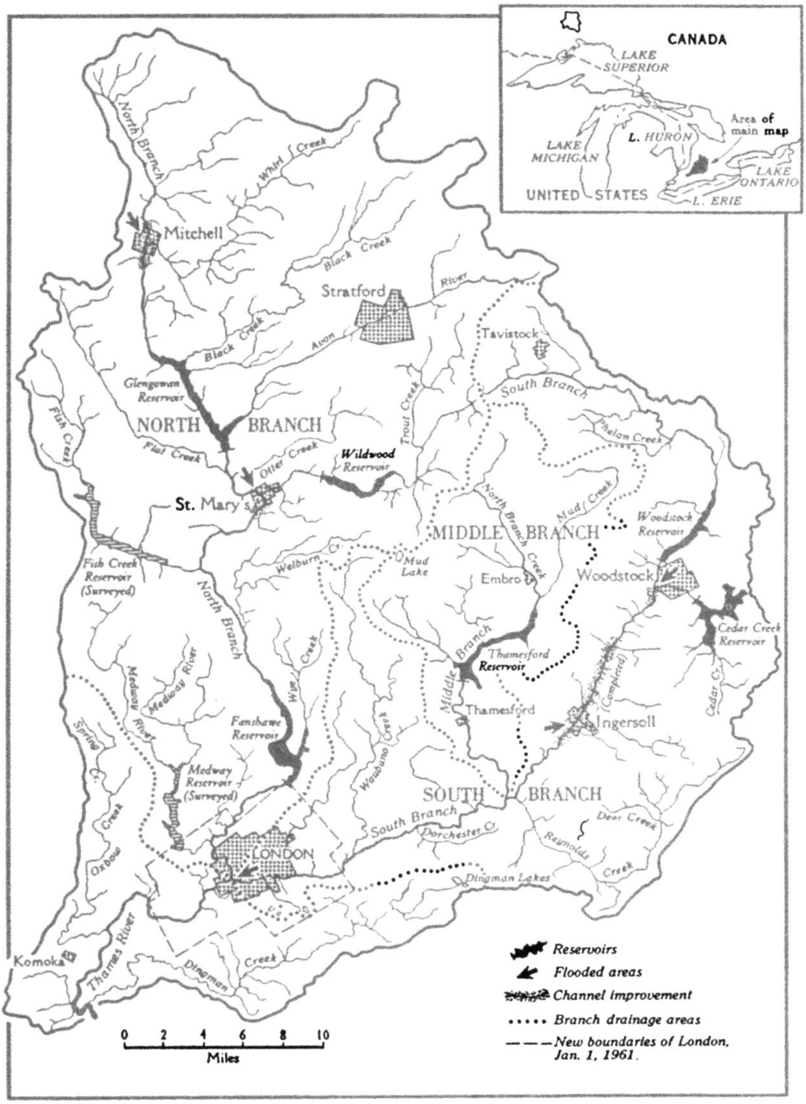

UPPER THAMES RIVER WATERSHED. Only the relatively large-scale water development projects can be located on the map. A number of other of the Authority's land and water activities are carried out widely throughout the watershed.

through the Authority 25 per cent. Fanshawe Dam, together with Fanshawe Park, have become a kind of prototype for similar projects elsewhere in Ontario. The permanent lake extends back from the dam over four miles and has a width averaging one-quarter of a mile with a total area of 650 acres. The ten miles of shoreline and 1,600 acres of fringe land owned by the Authority are available to the public for recreation. Motor boats of all types, including outboard motors, are prohibited on the lake; as a consequence the sailing of small boats, the organization of rowing clubs, and fishing have become important on Fanshawe Lake.

Most of the park acreage at Fanshawe Lake is in reforestation, nature preserve, and various development stages under the resources programs of the Authority. It is impossible to predict what future generations will need in their space requirements for outdoor recreation. The obligation of the Authority is to acquire sufficient superlative land, to establish a program of natural reconstruction, and if necessary to put the area in a kind of mothballing to await the needs of future populations. The Authority is acquiring land without trying to justify an intensive use for each square inch.

The Upper Thames River Conservation Authority has undertaken many other projects besides the Fanshawe Dam as part of its blueprint of development. A number of small dams have been built, several channel improvement projects have been completed, millions of trees have been planted on public and private lands, hundreds of farm ponds have been planned, built, or subsidized on private lands, and scores of well-devised conservation education programs have been carried out. These programs included projects in pollution abatement, fish and wildlife management, land-use improvement, farm planning, soil conservation, windbreak planting, woodlot improvement, recreational land use, tours, contests, and demonstrations participated in by school children.

Representatives of the federal and provincial governments met at Fanshawe in January 1961 to sign the agreement providing Federal aid for the construction of nearly $10 million worth of projects to complete the river valley program begun fifteen years before. The completion of the program in ten years will put the Upper Thames drainage basin under complete water management, the first drainage basin in Canada to be developed on a conservation basis. Five dams and three channel improvement schemes are included in the project. By 1970 the remedial works will control floods and increase summer flow throughout the 1,325-square-mile drainage basin.

The experience of the Upper Thames River Conservation Authority may be summarized as follows:

1. The Conservation Authorities Act is an effective body of legislation to enable the people in a drainage basin to work together for a regional conservation program.
2. It is necessary to prepare as complete a survey and plan as your financial resources permit.
3. It is imperative that the report with its recommendations be placed in the hands of the people who are most concerned specifically with the future, namely the elected and appointed officials, the teachers in the elementary and secondary schools, and the residents of the area. All means of public information and education should be used to inform the public and to earn its support.
4. The timetable of projects must be realistic. It is impossible to get public acceptance for a complete program immediately. The first project should aim at a partial solution of the major regional problem and at the same time give promise of beneficial side effects. For example, the right first project for the Upper Thames Valley was Fanshawe Dam which lessened the flood hazard at London and at the same time offered the extra dividend of an excellent recreational site for residents of the area and their visitors.
5. Research and practice will modify the long-range plan as the program is developed. The Conservation Authorities Act, when given meaning through action by an Authority operating in a watershed, is an extremely important legal device to enable the citizens to work out

democratically and carry through to completion a comprehensive development program based on sound conservation theory and practice.

THE CONSERVATION AUTHORITIES ACT IS, AS WE HAVE SEEN, a self-contained piece of legislation under which a wide variety of land and water programs is being carried out. In addition, however, it is related, in some instances quite closely, to three other broad lines of public action that involve planning for the future. In conclusion, let us at least mention these further elements in Ontario's long-range management of natural resources.

1. The Planning Act, first passed in 1946 and since then extensively amended, enables a municipality to establish a planning board to direct the development of the community towards socially desired goals. The Minister of Municipal Affairs may designate several municipalities or parts of municipalities as a joint planning area. This Act has proved to be effective and over two hundred planning areas in Ontario have been established under it. Chief among the several responsibilities of a planning board are to prepare the official plan for its area and to forward the proposed plan to the council of each municipality involved. When adopted, the official plan indicates the desired pattern of development, and municipal zoning by-laws must be in accord with it.

2. The Water Resources Commission Act, passed in 1957, established the Ontario Water Resources Commission to deal with the problems related to water supply and sewage disposal. The Commission has broad powers in pollution abatement. The government recognized water resources as a regional problem that must be solved on a regional, not a local basis. The legislation permits the province to enter into agreements with local municipalities or combinations of municipalities for purposes of water supply and sewage facilities. The Commission arranges the financing of the projects, builds the facilities, operates the installations, and recovers the costs from the local authorities over an extended period of time. In 1961 the

Federal government enacted legislation whereby contributions may be made to the provinces for projects related to sewage disposal and pollution control. The Commission estimates that over $2.5 billion worth of projects will have been completed in the province by 1975.

3. Under the Highway Needs Program the Ontario government has prepared a twenty-year plan for highway construction. The plan, developed by the Minister of Highways and his department, has two related parts: the first deals with the provincial highway grid of major thoroughfares and the second deals with local roads and the roads and streets within the municipalities. In effect, the official plan of the province indicates the general framework of development. It will not give specific siting for a highway but will indicate clearly that a limited access dual highway shall be built between two traffic-generating areas. The local planning groups, however, are given considerable information of the evolving road pattern.

By way of example, let us glance briefly at how London, Ontario, is using the three types of program. (Because the city is located within the Upper Thames River Conservation Authority it is not necessary to make further note here of how the Conservation Authorities Act is applied in the area.) London is a central city with a population of 162,000. The area of the city was increased from 11 square miles to 67 square miles on January 1, 1961. The Ontario Municipal Board, after an extended series of hearings on the city's application to annex portions of surrounding townships, ruled that the urban area is better administered and planned as a unitary municipality including sufficient land to give complete control over the urban fringe. The official plan, drawn under the authority of the Planning Act, recognizes agriculture as a legitimate land use and makes indirect provision for its protection against the premature blighting of urbanism. For instance, to control fringe development, the minimum size of lot is 25 acres in the development zone immediately around the built-up part of the city. In the agricultural zone farther away from the city the minimum size of lot is 5 acres. The official plan also states that urban development shall take place only when municipal water supply, sanitary sewers, and complete sewage disposal are available. And then, urban development may proceed only by a registered plan of subdivision.

The city of London now has three sewage disposal plants with four in process of construction. The new plants are being built by the Ontario Water Resources Commission and will be paid for by the municipality in 35 years. At least two more disposal plants in London are in the planning stage.

Up to the present time the basal gravels have given London an excellent water supply from deep wells in and immediately around the city, but increased per capita use and an increased population indicate that alternative sources of supply must be developed. Plans being worked out by the city and Water Resources Commission call for a 50-mile pipeline from Lake Huron to supply the London area with water of unusual quality in abundance. (It is a commentary of our times that the designation "unusual quality" means simply *good* water.)

The London area has an official plan of major thoroughfares which reflects the provincial planning as applied to a locality.

Thus a given urban district such as London is able to plan its future through the use of legislation devised to enable the citizens through their elected councils to steer the elements of growth and change towards socially desired goals.

There is no conflict in the application of the four kinds of legislation to the problems of a specific area or region. The Planning Act and the Conservation Authorities Act enable local municipalities to organize for the solution of problems that are related to day-to-day orderly development in fixed locations. The use of the legislation depends primarily on the initiative of the local municipalities and is adaptive in nature. The municipalities already firmly established in local government join together in joint planning and operational devices for the purposes of regional or area urban development and conservation projects and programs.

The Water Resources Commission Act and the Highway Needs are related to circulation: in the one case the supply of water and its subsequent treatment after use, and in the other the movement of goods and people. Provincial initiative is necessary in dealing with the dynamic problems of this kind.

Thus any area or region may use all the legislation without conflict. For example, a given urban center will join with its surrounding municipalities under the Planning Act to control urban growth; it will join with all the other municipalities within its drainage basin under the Conservation Authorities Act to promote the conservation of regional resources; it will enter into agreements with the province for the financing, construction, and operation of water supply systems and sewage disposal plants; it will relate its program of street and road construction and maintenance to the official plan of roads through the grants structure.

Planning is essentially a dynamic process in which a democratic people may decide the kind of a future it wishes to have. But planning requires legal sanction in order to be effective. Ontario has enacted some effective planning-sensitive legislation and the people of the province are learning to use it.

WHAT THE UNITED STATES CAN LEARN

■ AYERS BRINSER

THE OPPORTUNITY TO OBSERVE the several tactics and strategies used by nations comparable to the United States to develop their natural endowments is, in my opinion, an occasion of considerable significance. This is a field of inquiry that has too long been neglected. While "economic development" has become in academic circles a most fashionable topic of discussion, there has been little serious effort to compare the systems of resource development abroad with those at home to gain useful insights into how our own methods might be improved.

■

AYERS BRINSER is director of a study of the land-grant colleges of New England for the New England Board of Higher Education, and is also visiting professor at the University of Michigan. During the 1960-61 year he established and directed at the University of Colorado the only graduate study program in administration and policy of natural resources in the United States. He also taught advanced courses in the University's economics department. From 1952 to 1960 he was a faculty member in the Graduate School of Public Administration and director of the Land Use and Conservation Seminar at Harvard. Earlier, he was a dairy farmer, a junior editor for Time and Fortune magazines, editor of Every Week Magazine, and an editor for Harper Brothers. From 1941 to 1949 he served with the National Resources Planning Board, the Office of Price Administration, the Federal Reserve Bank of Boston, and the New England Regional Milk Price Committee. Mr. Brinser was born at Steelton, Pennsylvania, in 1909. He received his A.B. and Ph.D. degrees from Harvard University.

This myopia cannot be dismissed simply as pathologic chauvinism. It is quite possible to justify the parochial approach to resource management policy on the grounds that our concepts of policies and programs are derived from our own landscape and the processes of history that have evolved in that landscape. Thus comparisons, if they are to be useful, must be something more than a compilation of codes, manuals of administrative behavior, and budgets. Instead, they must be addressed to an investigation of the underlying concepts of social means and ends that are the determinants of differences for both policies and programs.

It seems to me that Dr. Pleva's paper provides ample ammunition for this line of attack. He has given a wide array of specific data about the functioning of the resource management authorities in Ontario. What is more, he has outlined the rationale of planning and administration. The opportunity to comment on these details is very tempting. To do so would also avoid the formidable problem of reducing to an orderly discussion the very complex, vague, and often conflicting concepts out of which grow resource policies and programs. Unfortunately, this is an insufficient excuse for ignoring the obvious fact that a meaningful comparison of resource development in the United States with that in Canada must be based on an examination of these concepts.

If a distinction can be made between the conceptual implications of the Canadian approach to resource policy and that of the United States, then it should be possible to establish the basis for real differences in tactics and strategy. The contention here is that this distinction is to be found in the consensus about the objectives of resource policy and programs. The core of the consensus in Ontario is a commitment to maximize net social gains. This takes precedence over the corollary commitment, to minimize social costs. The criteria for allocating the goods and services produced by public resource development programs with this objective do not rest primarily on the prices generated in the market. In other words, the preservation of a free market is not in itself an objective of resource development policy. Here allocation and distribution are guided by a combination of market prices and political decision with the latter frequently playing the larger role. This position is indicated, but incompletely expressed, in economic terms as the maximization of external economies and the minimization of external diseconomies.

In the United States, on the other hand, the major commitment is to achieve a maximum efficiency in allocating resources within the framework of a mixed market economy. This does not imply that social values are automatically sacrificed on the altar of economic efficiency; witness the growing significance of secondary and off-site benefits in benefit-cost calculations. Nevertheless, the emphasis is on an efficient market calculus as the most important discriminating device.

When we in the United States observe external economies and diseconomies we seek to reduce them to an equilibrium within the framework of market efficiency. Indeed, we tend to look upon these economies and diseconomies as aberrations of the desirable market equilibrium rather than as opportunities for making net gains. From this point of view, government is obligated to adjust these "disturbances" so that the gains and losses will be distributed in a way consistent with some assumed level of market equilibrium. This we usually do by taxation and subsidy. We do not seek to create external economies or prevent external diseconomies as a device for maximizing net social gains by public action.

It should be obvious that the reason for adopting the concept of the market as the essential equilibrating force is to avoid setting up normative values for political objectives. On the other hand, in the area described by Dr. Pleva, the necessity to deal with such values directly is accepted as a central obligation by the resource planning agencies.

In large measure, this distinction in economic consensus follows from critical differences in the political consensus. This is most clearly shown in the contrasting roles of the states and provinces vis-à-vis the federal governments. In Canada the more positive role in the management of resources is taken by the province while in the United States the federal government is the more active agency. This is true despite the fact that in the United States powers not explicitly assigned to the federal government are assumed by the states while in Canada the situation is reversed.

A further political distinction is revealed in Dr. Pleva's discussion of Ontario. The Province has a relation to Canada as a nation quite different from that of any state in the United States. Ontario represents a much greater proportion of the total economic power and energy of Canada than does any state in the United States. In the direction of resource development in Ontario a large segment of the total Canadian economy is affected. Because of Ontario's great relative weight, the authorities there are comparatively free from the controls and forced adjustments to conflicts of interest from other areas. In the United States, on the other hand, there is a vast array of energies, interests both social and economic, regions and pressure groups with sufficient power to force compromises, or at least place constraining conformities on national programs and policies.

Finally, there is in Ontario the inheritance from England, a respect for the landscape and its aesthetic qualities. In the English tradition, the earth is a garden to be kept and enjoyed by man. It is something more than a resource base upon which to build economic prosperity. Clearly, the Authorities Dr. Pleva has described have not been slow to exploit this consensus in planning for the enhancement of the aesthetic values of the landscape.

This political consensus is reflected in the administrative approach to resource management. In Ontario more direct power is delegated to the bureaucracy than is the practice in the United States. Protection of individual rights and provision for public assistance is embodied in general legislation with the result that the test that these be in the public interest is unavoidable, clear, and specific. In the area of resource management and development in the United States these rights and this assistance are usually provided by specific laws to serve identified interest groups such as irrigation farmers or miners.

In Ontario these protections take several forms. First, there is the party responsibility to programs which tends to enforce a general criterion for the equitable operation of governmental activities. The measure of equity is the effect upon the total society rather than upon defined clientele groups or individuals. It is not a case of the farmer versus labor or the city, but rather adjustments toward a total benefit to society. Policy formulation and the planning and execution of policies rests in the hands of representatives of the total community, as the elected members of municipal boards rather than in an elected body representing a special interest. Finally, these protections are reinforced by a sustained educational program for the whole community. This is aimed at informing the people of the specific program and policy alternatives at hand for reaching given social objectives.

THE ESSENTIAL POINT OF THIS DISCUSSION IS THAT WHILE THE United States and Canada hold to very similar values about the nature of democratic government, there are significant differences in the values placed on the processes by which the primary values of democratic government are realized.

A critical result flowing from the concepts that shape the approach to resource management in Ontario is that the focal point for policy formulation, program design, and authority is the resource region. This is in marked contrast to the situation in the United States where policy and program are centered in specific action agencies such as the Bureau of Reclamation or the Soil Conservation Service. Authority is thus divided between local interest groups, state agencies, and federal agencies, with the predictable result that one of these is forced by circumstances to play the dominant part. The problem area is circumscribed by the program of the dominant agencies and the solutions they are empowered to provide.

To establish a region as a focal point, the first obvious requirement is that an agency based on the region have the primary authority for planning. The identification of the problems to be solved, the order of their precedence, and how they are to be managed are defined by the total objective to be achieved in the region. The role of the provincial and federal governments is to provide expert advice and analysis, and marginal funds.

In this context the Conservation Authorities in Ontario have a real basis for autonomy. This is reinforced by the budgetary powers given to the Authorities. They can raise their own essential funds, and spend them to solve the identified problems. Equally important is the executive power vested in the Authorities to carry out the planned programs. It is indeed a formidable array of powers that Dr. Pleva has outlined. They include the taking of land, the spending of public money to hold land and water in reserve for predicted future public use, the setting of payment for land purchase on the basis of future estimated social value rather than market prices, the allocation of land to uses that would optimize net social gain in spite of high opportunity cost in alternative private use. While we in the United States have on occasion put on paper our intentions to deal firmly with the regulation of private land use, we have rarely had the temerity to use those powers with the bold assurance that would seem to be common in Ontario.

This assurance rests apparently on the concept that the objective of resource development in Ontario is social gain to be made available directly to society over time rather than to individuals through a subsidized market procedure. The whole community pays the costs and receives the benefits which are distributed on a per capita basis. This perhaps explains why in Ontario subsidies to individuals for such practices as erosion control and water management are so much smaller than those handed out in the United States.

An obvious advantage of centering a resource management program on a region is that the alternative solutions are automatically taken into account. The decision about what the central problem is, and what the available solutions may be, does not depend on the more narrow objectives of the programs offered by special-purpose, outside agencies. These, rather, are determined by an examination of the defined problem area. From this scrutiny a policy is established, and then programs devised to implement that policy. The problems of watershed management afford an example. Under the regional authority the first issue is not whether to adopt a system of upstream engineering, or major stream control works, or a flood warning system, or flood plain zoning. Planning and policy are instead aimed at optimizing the return from development investment co-ordinated to provide flood control by the most effective method consistent with maintaining needed water supply, recreation, and future space requirements.

These clear advantages must be weighed against certain apparent disadvantages. When the primary focus of policy is on the generality—to maximize net social gain—there is the possibility that useful efficiency criteria may be given short shrift. In one of the several Ontario Conservation Authority plans that I have seen there is the statement that benefit-cost calculations should not be permitted to interfere with responsible decision making. Most readers, I am sure, are aware of the weakness and perversions of the benefit-cost ratio as it is used in resource development planning in this country. On the other hand, the mere fact that it is not a precise indicator for investment fund allocation is scarcely sufficient reason to make a burlesque of it. Surely a rational society is interested in the net cost of alternative possibilities in an ordinal if not a cardinal scale. One does not have to accept all of the economics of Poor Richard to agree that paying too much for one's whistle does limit the range of desirable choice.

The problem here is illustrated in the analysis made of recreation benefits on a Toronto watershed. Having dismissed the benefit-cost ratio as an impediment to responsibility, the planners went on to concoct such a ratio, or a reasonable facsimile. The estimates of recreation benefits came to six times those from flood protection. The astronomical (and unsupported) value assigned to recreation was necessary to produce an economic justification for the total investment. Perhaps recreation deserved this weight of value, and perhaps also, the only way that value could be stated was as an educated intuition. But to call that intuition a quantified value in a benefit-cost calculation did little more than prove the known fact that the benefit-cost ratio can be reduced to meaninglessness.

When a planning area is defined by physiographic features alone, such as a river basin, there is a tendency to force the consideration of alternatives in terms of physical rather than economic or social consequences. Planning is aimed at maximizing what are said to be physical qualities of resources rather than their economic or social values. The plans for Ontario which I have read seem to me to have overemphasized the prospective returns in physical quantities and qualities of timber, soil, and water rather than the productive yield from these resources in use.

To measure the results of investment planning in terms of the physical quality of the product usually limits the alternative possibilities to fixed and inflexible programs. Physical features at any one time appear to be immutable. Certain values are assigned to their physical quality, as, for example, to produce a long, clear, butt log becomes the highest objective of forest investment regardless of predicted technological changes in manufacture or the product market. Thus planning for future forest production is set in terms not of the use of the product, but rather in terms of the physical properties of the raw material. In certain quarters these physical values become moral imperatives.

There is another side to this question of flexibility. I have paid my respects to a resource planning system that forces a consideration of all alternative means of solving a given problem. In the United States we are familiar with the problem arising from the lack of unity in our approach to resource management problems; one might be tempted to say, lack of policy. In our efforts to deal with the conflicts and duplication created by various agencies attempting to promote their programs as total solutions to partially defined problems we have written a great many memoranda of understanding and prescriptions directing how everyone, from the field to Washington, will work together for the common good. It would be ungenerous to attempt to measure the weight of their impact.

Instead of starting with an integrated plan, we approach the problems as they are observed with a variety of plans and programs and then attempt to co-ordinate the work of the agencies concerned after the work has begun. In addition we have divided budgetary authority, divided executive control, and a not inconsiderable rivalry to see who will get to the fire first.

Many observers, including the members of the task forces of the Hoover Commissions, when they contemplate this disorder, find it imperative to prescribe an administrative cathartic. On the other hand, given the demonstrated fallibility of predictions of future needs, the great importance of flexibility of controls, the variety of forces and pressures in the United States, competitive programs and competitive approaches can be very creative. No small amount of most useful innovation has grown out of this scramble that is so frequently and so sincerely deplored.

Dr. Pleva has indicated the efficiency of management made possible by the relatively tightly knit structure of administrative authority in Ontario. In the United States, many people would regard this degree of control to be a threat to democratic decision. In our resource management programs great emphasis has been given to local elected authorities representing the people affected. If one were wholly candid he would have to admit that the success of many of our resource programs depends on the way in which the program agencies have led these so-called grass roots leaders by the hand to the point of taking action.

Resource management programs that impinge on the private management of land of necessity require a degree of social control even though this may be disguised under a blanket of subsidies. In the United States as well as in Ontario there is an inescapable demand to manage resources so as to provide some level of net social benefit. The real issue here is not only to identify the objectives of policy, but to find the means of carrying it out that will be consistent with the objectives and the accepted forms of political control as well.

The methods of control used in the United States give great weight to clientele interests. This objective apparently is not held in such high esteem in Ontario. But regardless of the commitment to the "grass roots" approach in the United States, we are being forced to recognize that the combination of conventional market criterion and local control is not a sufficient formula for an effective resource management program. Whether, given the complexity of the United States economy it would be possible to impose as rigorous control, or to ignore group interests to the extent that this is apparently done in Ontario, is dubious. Nevertheless, it is clear that many of our attempts to be responsive to group interests are self-defeating not only when measured by net social gain but for the clientele groups as well. The experience in Ontario suggests the direction some adjustments in our approach might take.

For a long time the major difficulty in resource management policy in the United States has been to secure real participation by local and state authorities in nationally sponsored programs. As was indicated earlier, this has been achieved with considerable success in Ontario. The price we have paid for our failure here has been higher cost caused by investment that returns less than could have been gained from alternative opportunities. We have used some programs to meet objectives which could have been reached more efficiently by other means. There are certain unproductive selling costs when one agency competes with another in an attempt to have its program adopted over those of its rivals. We have often spoken of the state as the laboratory of social invention, but that is an opportunity that we have seized all too rarely.

Nevertheless, given all of its faults, our approach in the United States has been reasonably flexible. It has led to important innovations. We have bought these advantages at considerable cost, but not necessarily too high a cost. Our emphasis on benefit-cost quantification, even though *ad nauseam,* and sometimes to no particular purpose, has at least been educative even if it has given us no precise tool for discriminating. It is important that the voting public should know that the distribution of costs and benefits does affect the net value of social benefit produced. The final explicit decision in resource management is a budget decision, and some form of quantification must be incorporated in the budget if it is to make sense.

Clearly there is no answer as to which of the two approaches to resource management is the better. I believe that the method used in Ontario deals with the normative issues in a more forthright and effective way than does ours. We are at least becoming increasingly aware of the importance of this issue. Our difficulty has been to give precise definitions to these norms, and to face them as factors to be incorporated in our analysis of policy rather than exclude them simply because they are norms. In my opinion, the major unresolved issue in resource policy today, in both Ontario and the United States, is not the maximization of the quantity of output as measured by prices paid, but rather one of improving the quality and distribution of the product. I think Dr. Pleva has indicated how Ontario has made real progress in this direction. This we could most profitably observe to our own benefit.

VI FRANCE:

DEVELOPMENT OF A REGION

JOHN H. NIXON

The Nature of the U. S. Problem

PHILIPPE LAMOUR

● *Land and Water Development in Southern France*

STEFAN H. ROBOCK

What the United States Can Learn

1. DIVERSION WORKS ON THE RHÔNE NORTH OF ARLES.

2. THE MAIN PUMPING STATION. This installation, located near Saint-Gilles, is the largest in Europe.

3. **THE PRINCIPAL CANAL,** between the Rhône and the pumping station near Saint-Gilles.

4. **A MAIN BRANCH CANAL,** with its secondary pumping plant and water tower, that supply the underground irrigation network.

5. TRADITIONAL CROP. A typical vineyard landscape.

6 & 7. NEW CROPS. Left, orchards irrigated by sprinkler system; Right, rice lands.

8. THE MODERN FARM MARKET AND STORAGE CENTER NEAR NIMES.

9. RURAL HOUSING. One of the new homes built by the National Company for Development of the Lower Rhône-Languedoc Area.

THE NATURE OF THE U. S. PROBLEM

■ JOHN H. NIXON

AS A SETTING FOR PHILIPPE LAMOUR'S discussion of a most interesting resource development program in Southern France, I shall review briefly some of the problems the United States has encountered in regional resource development. Since the history of the European settler on this continent stretches back several centuries, one can recall a good number of development problems. A seamless web of events and policies connects the past with the present, and I find a backward glance. useful in establishing our present orientation.

From the beginning of recorded history man has been struggling to maintain a proper balance between himself, the soil, and his supply of fresh water. Economic abundance has been associated with effective use of these resources; destruction of soil or water supply through ignorance or misuse has

■

JOHN H. NIXON is director of the Area Development Division of the Committee for Economic Development (CED). From 1956 to 1959 he was Director of Economic Development in the New York State Department of Commerce. He has been a consultant to the President's Materials Policy Commission, the U.S. Office of Price Stabilization, The Brookings Institution, the Center for International Studies at M.I.T., and for private industry. Earlier he served as an economist with the U.S. Office of Price Administration and the U.S. Railroad Retirement Board in Washington, and he has taught economics at Harvard College and The City College of New York. Mr. Nixon was born in Minneapolis, Minnesota, in 1915. He received his A.B. degree from Swarthmore and his Ph.D. in economics from Harvard University.

brought decline to regions and cultures. Productivity of resources has been dependent not only on technology but also on the social institutions men use to guide their economic behavior.

For centuries before the white settler came, the North American continent had supported an Indian population which maintained an effective balance between itself and the soil and water resources. Their primitive technology and limited social institutions, however, placed severe restrictions on the size of the population that could be supported.

The European colonist brought both a more advanced technology and a set of social and political institutions that enabled thousands and hundreds of thousands of people to co-operate via specialization and exchange. The industrial revolution of the eighteenth century and the rapid development of technology in the nineteenth, both in Europe and America, made it possible for the North American continent, and particularly the area of the United States, to support a real population explosion.

Regional resource development in the United States originally meant the displacement of the Indian by the settler. As the pressure for westward migration grew, the provision of adequate transportation routes became a fixed purpose of government policy. The states early assumed responsibility for construction, or for financial support of private construction, of turnpikes and canals and eventually railroads.

As a young legislator in Illinois, then a new frontier, Abraham Lincoln supported a bill to provide state financing for a system of railroads. canals and turnpikes which would connect most of the settlements in Illinois with each other and with the outer world. Later, as President, he signed acts which made possible the financing of a railroad connecting Omaha and the Midwest with San Francisco. Twenty sections of public land were given to the railroads for each mile of track laid, and substantial loans were made to the railroads, secured by second mortgage bonds.

Adequate transportation has always been an essential part of regional development, and government at the federal, state, or local level has accepted the final responsibility for providing transportation facilities according to preferred technology. Private enterprise has generally been encouraged to finance and provide as much of the roadbed and of the vehicular service as possible. Since the advent of the automobile there has been practically no private construction of roadbed, though borrowing by toll road authorities and similar public bodies has been an important financing device.

The development of the soil was left to individual initiative. The government sold public land to settlers at low prices in the early part of the nineteenth century and eventually offered it free to homesteaders under the Act of 1862. In the same year a federal Department of Agriculture was established and the Morrill Act, which started the land-grant colleges, was passed. Since then scientific research in agriculture and the training of farmers in the most advanced technology have been accepted functions of government. Later these policies were further implemented by the development of state agricultural experiment stations and of the county agent system, and by provision in rural areas of agricultural vocational education in secondary schools. More recently the Rural Electrification Administration and the Soil Conservation Service, among other agencies, have helped the individual farmer to apply modern technology to the soil. We must not neglect, however, the great contribution which private and industrial research have made in developing labor-saving machinery.

The lack of water has prevented development in many parts of the arid West; in other sections the waste of water has meant flood, loss of topsoil, and distress. The technology of the twentieth century has made possible a multiple-use of river basins if developed according to an integrated plan. Navigation, flood control, irrigation, hydroelectric power for industrial and domestic use, and recreation are among the purposes which can be served.

Accompanying these policies towards transportation, land and water has been a general policy of maintaining a dynamic, competitive, free-enterprise system in manufacturing, trade and services. Although this policy has no regional focus the multitude of free decisions made by investors, managers and workers has been a dominant force in regional development and change.

The shifts in regional specialization and in regional population distribution since 1870, and particularly since 1939 have recently been described and explained by Harvey Perloff and his associates in *Regions, Resources, and Economic Growth,* published in 1960 by The Johns Hopkins Press.

We have seen a shift in major industries such as the movement of the textile industry from New England and the Mohawk Valley of New York to the South, a decline in the share of automobile manufacture in the Detroit area, the shift in major fuel use from coal to oil with serious impact on employment opportunities in Pennsylvania, West Virginia, Kentucky and Illinois. Yet at the same time we have seen the rapid growth of the electronics industry in a New England deserted by textiles and the growth of consumer goods manufacturing in the Los Angeles area alongside of aircraft and missile construction.

The loss of industry, awareness of the possibilities of attracting new industry, and the desire for growth, have all contributed to the development of many public and private organizations devoted to promoting the economic development of particular regions and localities. A study by the Committee for Economic Development reported the existence of 14,000 local and regional organizations, each working to help its own region or locality, and generally competing with one another. The number continues to grow apace.

Effective resource development at the regional level usually passes, I suspect, through a series of stages.

First, there are some preconditions to development. A gap must exist between the technology by which the resource is currently utilized and the best existing technology which could be used. Then there must be an economic need or incentive to close the gap. Perhaps the area may have lost an industry and have workers who need jobs; perhaps new methods of exploiting the area's resources offer profit possibilities that make the application of technology clearly worthwhile. And someone must have a vision of the benefits which would accrue from closing this technological gap. These visions often vary, and many turn out to be only pipedreams; still, without the vision there would be no action.

Second, a broad strategy must be worked out for applying modern technology. This strategy includes a general assignment of administrative responsibilities among the central government, local government, corporate enterprise, and individual effort. To take the St. Lawrence Seaway as an example, a corporation was formed under the auspices of two national governments—the United States and Canada—to build the seaway itself. Industrial development work has been left to various state governments and localities; improvement of the rivers and harbors along the seaway is the province of the Corps of Engineers. The actual use of the power developed at Messina (half of which is going into aluminum reduction) is largely in private hands. Most of the employment effects in the area come out of this final private effort. The strategy includes also a division of the burden of financing. An allocation among public taxation, public borrowing, private capital, and private borrowing must be made. The time horizons involved, the possibility of charging for direct benefits of a development project, the income and tax yields of indirect benefits of the project all enter into the planning of a financing program. It is in the area of broad strategy that most of the debate on appropriate public policy centers. Here political philosophies clash. Individual or group concerns for self-interest come in conflict, and the differences in expectation of ultimate benefits create controversy.

Third, within the lines laid down by the broad strategy, a detailed program must be administered. Policies and programs must in some fashion deal with the people already in the area; with land, water, transportation and power sources; and with a wide range of normal or possible economic activities. The difficulties to be faced by government administrators, private entrepreneurs and just plain people at this stage are so great that I will not mention them, but merely express thanks for human fortitude.

This somewhat meandering introduction has covered a variety of aspects of American economic policy. This, perhaps, is the most significant thing about it. We have never attempted in this country to combine all of the elements that have been touched upon into one great effort to change the economy of a particular region. Our programs have operated singly or in limited combinations.

Our broad strategy has been to act as a mixed economy, with private enterprise responsible for the major share of vision and effort. The allocation between public and private effort shifts somewhat with circumstances. I am sure we can benefit from the experience of our neighbor in the Western community.

LAND AND WATER DEVELOPMENT IN SOUTHERN FRANCE

■ PHILIPPE LAMOUR

DURING THE PAST TEN YEARS a far-reaching effort has been under way to improve the use of land and water resources in the Lower Rhône-Languedoc area of Southern France, where traditional agriculture is in distress even though its potential, under modern methods, is great. Conceived as an integral part of a plan for the whole nation, the program nevertheless takes special account of the problems of a particu-

■

PHILIPPE LAMOUR is president of The National Company for the Management of the Lower Rhône and Languedoc Region (Compagnie Nationale d'Aménagement de la Région du Bas-Rhône et du Languedoc). He also is president of the European Committee for Underdeveloped Regions, vice-president of the European Productivity Agency, member of the French General Planning Commission and of the National Productivity Council, and president of the National Federation of Superior Wines. He was born in 1903 at Landrecies, a small town in a farming district of northern France, studied at the University of Paris, and practiced law until 1939. After serving in the Second World War, he settled as a landowner near Nîmes in the south of France, and has served as secretary-general of the National Farmers Association. Along with farming, he has become interested in national and regional economic planning.

Mr. Lamour wrote his essay in French. The text used here was translated by his associate, R. Corbière, secretary-general of the Lower Rhône-Languedoc Company.

lar region and aims at improving the levels of living of its people. It employs the latest advances of modern technology and is administered by a mixed company that brings together public and private interests. The program embodies many characteristics that are new for France and some, I believe, that will be novel to other countries as well. Especially for the latter reason I hope that this account of our experience will be interesting and useful to readers in the United States.

Because the Lower Rhône-Languedoc program, though distinctive in its operation, is part of a national program of economic and social development, it is necessary to begin with at least a brief look at the broader background.

France is a highly centralized country. Her political institutions, shaped by the French Revolution and the Napoleonic Empire, have resulted in complete administrative centralization. Even today any decision of importance is taken in Paris.

To be sure the country with a total area of 212,000 square miles and a population approaching 44 million inhabitants, is divided into 90 *départments,* each one supervised by a *préfet* representing the government. But in contrast to the federal form of government, the departmental system restricts the powers of the local authorities to applying government policy or making decisions of limited importance. The nationalization of important sectors of the economy (railways, electricity, coal mines, banks, etc.) during the last decades has further tended to increase this centralized character of France's economic, administrative, and political life.

However, for some years, there has been a tendency towards decentralization. It has become more and more obvious that the *départment* was an undersized unit, its average area covering no more than 2,000 square miles and its population averaging 500,000. By now, the public authorities understand that political decentralization must operate in larger areas than departments; their number should not exceed half a score.

This change in thinking did not, however, originate in politics but in recognition of a serious lack of economic balance. The northern part of the country includes the largest and most fertile plains, numerous and regular watercourses and enjoys a relatively well-balanced climate. Consequently, it has the more highly developed agriculture and most of the nation's navigable canals, roads, and processing industries. Also, two trends that began in the nineteenth century have had a decisive influence. As a result of the "concentration mania" of the ideologists and leaders of the Third Republic, nearly all the best equipped and most important railway lines have their terminals in Paris. And, in France as elsewhere, the industrial revolution in the middle of the nineteenth century was based on coal and iron ore, which are found in the north and east of the country.

France with her overdeveloped metropolis tends to possess but one large city up to modern world standards. Nearly one-fifth of the population lives in the Paris area. Two-thirds of the country's population and four-fifths of its income are concentrated in that third of the country that lies north of a Le Havre-Grenoble line. South of this line, average per capita income is half of what it is in the north.

In most cases, the displacing of factories and the creation of new establishments resulting from the industrial decentralization policy since the June 1955 Decrees have occurred north of that Le Havre-Grenoble line. That is why one must take a new look at the economic development and utilization of land resources in the light of two factors which are likely to dominate France's national life for the next few decades: population increase and adherence to the European Economic Community.

According to all the demographic evidence, France will in less than 20 years have 55 million inhabitants. This may even be a conservative estimate, as some of the French people now living in overseas territories are coming back, and the principle of free movement for persons inside the E.E.C. opens up the country to immigration.

One million new jobs are necessary before 1965 and two million before 1970. But it is not in the long-range interests of the nation that these jobs be located in the part of the country that already enjoys a fuller development even though investments there would have the best multiplier effect. They must be distributed among all areas where efficacious economic development is possible. Failing intervention in that direction there is a clear risk that economic development and the creation of new jobs will be reserved to the twenty or so northern departments that are already ahead, and even more so to the Rhine basin.

This does not mean that we want to scatter the economic development indiscriminately throughout the nation, or to concentrate industrial and agricultural activities exclusively in certain areas. We want to choose the most favorable areas so that our programs will yield the highest benefits. Except for Austria and Greece, France is the only country in Europe to have just one great city. We want to achieve the kind of situation that obtains in Germany, Italy, and Great Britain by creating some seven or eight large metropolises, each of which would be an intellectual, economic, and industrial center. Together they could balance the present excessive concentration in Paris. In the long run the existence in France of a city like Paris is enormously expensive to the nation.

At present, only northern and eastern France are pitched to European Economic Community standards. There, prosperity is to be found and always with cumulative effects. A proper balance between industrial and agricultural activities brings wealth and flexibility of employment.

The remainder of France, situated around the *Massif central* area, is turning its back on Europe's economic heart. Brittany is looking towards the Atlantic void (the United States is so far away!) ; the southwest towards the Spanish void. The Mediterranean southeast facing the east and Africa, where possibilities of expansion are enormous, offers the largest development potential. But its isolation must be broken. It needs to be reunited with the industries of northern Europe by creating a large artery of communication that through the Rhine and Rhône valleys will join the North Sea to the Mediterranean.

The different steps leading to economic development in

the south of France are best considered in a large political and economic perspective. The rational use of water and soil with the aim of improving the condition of agriculture through irrigation is of most immediate concern, then the development of hydroelectric energy and natural gas as long as nuclear energy production is not yet at hand (though it is more advanced in this area than in the other parts of the country), and finally the use of watercourses as means of transportation.

The hydrography of Southern France is dominated by a large waterway: the Rhône, which has a drainage area of 36,600 square miles. When it flows through Arles—an old Roman city on the Domitian Way, 20 miles from the Mediterranean—the Rhône has an average flow of 58,450 cubic feet per second, a normal flow at low water of 31,500 cubic feet per second, and an annual average volume of flow of 43.5 million acre-feet.

As a matter of fact, by European standards the Rhône is the only large river in France, and one that has been most spectacularly utilized. The development of the Rhône was actually launched in 1931 and sponsored by the Compagnie Nationale du Rhône in which President Franklin D. Roosevelt had shown personal interest during the New Deal. This company was entrusted with developing the Rhône in three respects: power generation, agriculture, and navigation. The main objectives were to build a number of hydroelectric plants between the Swiss border and the sea, with an installed capacity of 2,720,000 kw. and an average annual production of 14 billion kwh.; to construct locks in connection with every low-head plant; and to irrigate 250,000 acres in the middle Rhône valley. Four plants are now operating with an installed capacity of 1,055, 000 kw. and a production potential of more than 5 billion kwh.; two other plants are under construction. Navigation has made but little progress because the rapidly flowing stream has numerous difficult stretches. Only a broadly conceived policy that aims at establishing a large navigable waterway joining the north of France to the south could bring to full fruition the accomplishments and projects of the Compagnie Nationale du Rhône in the field of inland waterways.

The development of the Rhône basin has not been limited to the Rhône. More recently, its main tributary, the Durance, has also been developed, through large hydroelectric plants. The work was begun in 1950 and is today nearly finished. By now, it includes a dam with a storage capacity of 9.8 million acre-feet, a canal 56 miles long, six main power plants with a total of 885,000 kw. installed capacity and 3 billion kwh. annual production potential. At the same time, irrigation in the Provence has been improved on 250,000 acres and more than 125,000 acres are to be added.

Thus, long deprived of energy during the earlier period of industrial development, Southern France today has at her disposal in the lower Rhône valley the most powerful hydroelectric complex in Europe. The first consequences have not been long in following. Important energy-intensive industries have quite recently developed here: stainless steel works, ferroalloy plants, plutonium and enriched uranium establishments.

The south of France is equally endowed with other forms of energy: natural gas in the southwest and petroleum products in the southeast. In the southwest the gas field in the Lacq area, near Pau, had by 1960 reached a capacity of 245 billion cubic feet. Its methane content of 70 per cent is particularly high, about the same as contained in pipeline gas in this country. This new energy resource has allowed industry to expand in an area which had hitherto been rather poor: in the last few years eight important industrial centers have been created in the fields of petrochemicals, aluminum, sulfur, nitrogenous fertilizers, etc.

In the southeast oil refineries have reached a capacity of 14 million tons per year, or one-third of total French capacity. These refineries are located mostly near Marseille where they have contributed to the development of five large petrochemical units. Moreover, in a few years' time, North African natural gas will be piped to France and northern Europe via Spain. The southeast of France will then be one of the first areas to enjoy that new cheap source of energy.

Finally, the French government will, in the near future, have to make important decisions concerning the complete development of a navigable waterway joining the Mediterranean to the Ruhr region by way of the Rhône and Rhine valleys. This project, many portions of which have already been built in connection with the above-mentioned hydroelectric development of the Rhône, would contribute greatly to making the French Mediterranean basin an area whose economic level would be comparable to that of the northeast of France and the Rhineland.

For these reasons the economic future of southern France appears more favorable in 1961 than it did ten years ago. A whole complex of new factors and measures taken in favor of this area have contributed to this improvement. The development of the Lower Rhône-Languedoc region that is only beginning to bear fruits ranks first among those factors.

WHY WAS THIS PROJECT PLANNED? APART FROM GENERAL reasons which have contributed to intensifying the disparity between north and south, the economy of the Lower Rhône-Languedoc region had also its specific ills. To understand this particular problem, one has to bear in mind the history of some special aspects of the area's economy over nearly a century. French vineyards, as so many others, were nearly ruined in 1880 by infestation of plant lice, phylloxera. In the departments of Gard and Hérault (the two departments mainly involved in the Lower Rhône-Languedoc project) vineyards were largely destroyed and had to be restored. However, restoration did not occur on the same type of soils nor with the same type of plant. Most producers of French common wine left the hills for the plains; at the same time the American rootstock on which the French vines were grafted proved so highly productive that the acreage in vineyards has dropped from more than 6 million acres in 1874 to only 3.4 million acres, while average annual production has risen from 770 million gallons in the 1870's to 1,320 million gallons in the 1950's. To this must be added the continuously expanding Algerian wine production that today averages 375 million gallons.

In consequence of this situation, the production of common wine permanently exceeds demand, which refuses to expand with population growth. The dislike that the younger generation shows for alcoholic drinks is one reason for this stagnation. Another is that the export of ordinary wine is practically impossible, inside the Common Market area as well as outside, largely because the habit of daily wine-drinking is widespread only in the countries that themselves produce wine. Southern France, as the area which furnishes more than half the production of common wines, has been especially hard hit by the wine-growing crises. It has been estimated that between 1930, a relatively prosperous year, and 1951, a year of severe crisis, the real purchasing power of southern wine-growers dropped by 44 per cent. Many people moved from the impoverished Lower Rhône-Languedoc area; between 1882 and 1951 the cultivable land area in the departments of Gard and Hérault dropped from 935,000 acres to 325,000 acres, while the sheep population, formerly an important resource, dropped from 859,000 head to 226,000 during the same period.

Economic measures were taken by the government as early as 1936 to protect, regulate and limit wine production. New plantations were prohibited. But the wine production potential was already excessive and has barely decreased since. Thus, French wine-growing has remained in a state of permanent crisis save for a few years of exceptionally poor crops. Since 1936 the government has withdrawn part of the surplus wine production from the market to be distilled into alcohol. But there is no demand for this alcohol which costs ten times as much as synthetic ethyl alcohol. The government, which owns the alcohol, can sell it, but only at a price far below its cost. It cannot be sold abroad above the world market price. The difference between the price paid by the government to the producers and the possible selling price of the alcohol is charged to the Treasury, which during years of high production has lost as much as $40 million annually. Even this year, though crops have not been exceedingly large since World War II, surplus of ordinary wine reached over 500 million gallons.

Faced with so catastrophic a situation, the public authorities have more recently taken measures to reduce acreage devoted to wine-growing; these have been known as the policy of compensated voluntary uprooting. But this policy, though the most intelligent that has been launched in the field of wine-growing, is difficult to apply in southern France, where only grapevines, olive trees, and extensive grain cultivation can stand the very dry climate. That is why land not dedicated to wine-growing or olive trees is practically abandoned.

If new competitive farm products are to be substituted for wine-growing, water must be brought to the area. This is where the Lower Rhône-Languedoc irrigation project, the largest in Europe, has its real roots.

The idea of irrigation of the Languedoc is one hundred years old. But for various reasons, none of the numerous plans which had been drafted was ever successful. Not until 1951, when it was faced with the dangers and the cost of trying to protect wine-growing, did the government decide to take some serious interest in the problem. It was then that the General Planning Commission was asked to work out a co-ordinated plan for the development of the Lower Rhône-Languedoc area in the framework that included irrigation. The plan to be proposed to the government had to take into account all the consequences and economic advantages that would evolve from the plan's implementation.

The project concerns the Lower Languedoc plains situated along the Mediterranean, between the Rhône and the Pyrenees, an area that covers 625,000 acres, including the most fertile soil of the province. The population totals 650,000, half of it in five towns: Nîmes (100,000), Montpellier (100,000), Sète (40,000), Béziers (80,000), and Narbonne (30,000), and the other half in 230 villages. Wine-growing prevails and dominates the population's economic and social life, but about 250,000 acres are undercultivated because of the dry climate. The average annual rainfall of 20 to 30 inches might at first glance seem sufficient. But it is not evenly distributed; rainy days do not exceed 60 to 80 per year, mainly during spring and autumn, compared to 170 days in the north of France. On the other hand, the south enjoys far more sunshine: 2,700 hours per annum, and the average temperature per year is warmer by 6° F. than the median temperature in the country as a whole. Drought, then, is the only unfavorable feature in the Mediterranean climate; with irrigation, the other conditions set the stage for high yields.

The waters of both the Rhône and the coastal rivers can be used for irrigation. To the east the main canal, taking off just north of Arles, diverts 2,800 cubic feet per second from the Rhône. Pumping installations, of which the first one, located near Saint-Gilles, is the largest in Europe, will develop more than 100,000 horsepower. This will permit irrigation of 225,000 acres out of a total area of 325,000 acres. Because of the extremely stable flow of the Rhône, water supply does not appear to constitute a problem. Even when the long-range program is fully developed, diversion for irrigation would take less than a tenth of normal flow at low water.

In the western zone, 200,000 out of a total of 300,000 acres are irrigable. Irrigation water will come from two rivers, the Orb and Hérault, after the flow of these coastal rivers has been partly regularized by storage dams with an aggregate capacity of 250,000 acre-feet.

Besides irrigation it is planned to provide for flood control for the coastal rivers and for the drainage of the humid and sometimes salty flat lands along the coast; their very fertile soil makes their reclamation most desirable.

Within the 625,000 acres to be controlled under the plan, it is intended to switch 300,000 acres of vineyards that are high-yielding but produce poor quality wine to other kinds of production that are both more lucrative and more useful to the national economy. It will thus be possible to reduce the overabundant wine supply by an average of 155 to 175 million gallons a year. Moreover, irrigation will permit the improvement of 175,000 acres of good, though now underexploited or uncultivated, land.

The development plan for agriculture is one of the project's main features. Of course, it is essential to promote the most

THE LOWER RHÔNE-LANGUEDOC REGION

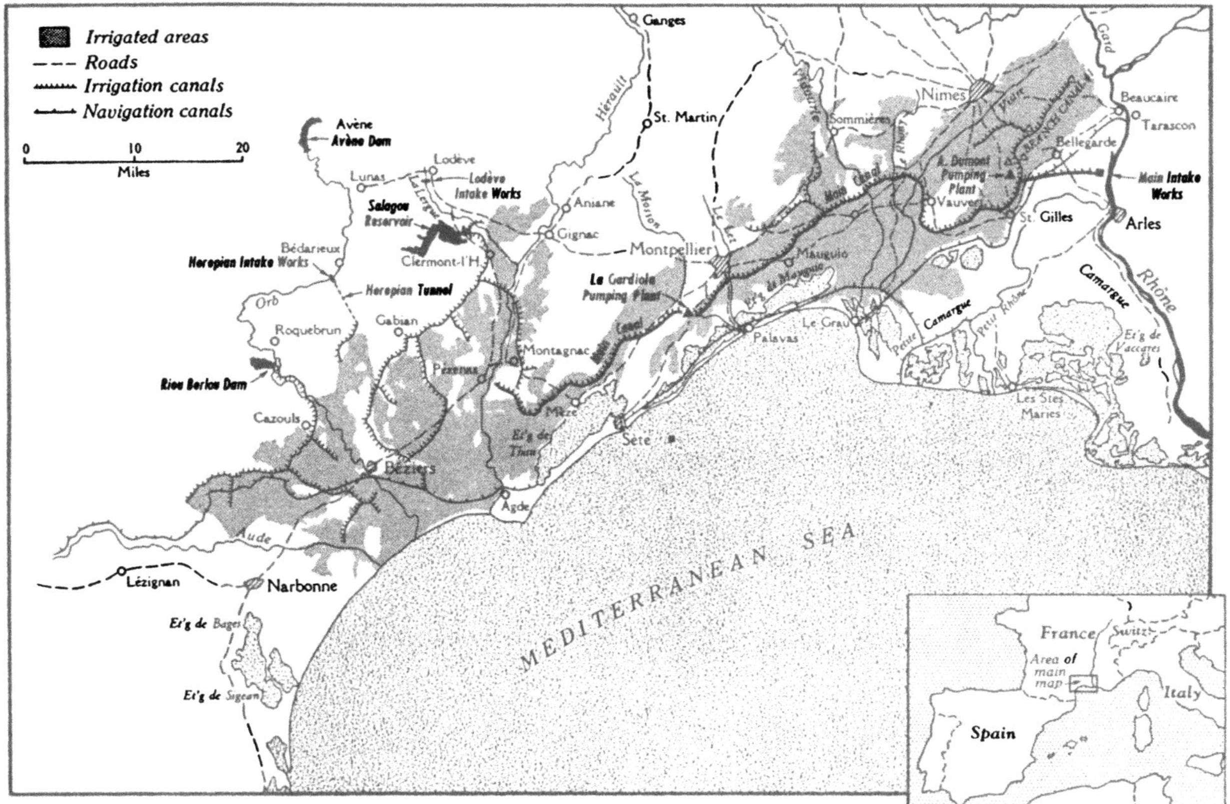

marketable crops and products and avoid those which are or will be faced with decreasing demand. In recent years, in France and in Europe generally, market studies have been undertaken to ascertain future food needs as a function of projected purchasing power. Price elasticity of food consumption has also been studied. The conclusions drawn from these investigations have made it possible to determine the orientation that agriculture should take; consumption of fruits, vegetables, poultry, dairy products, eggs, and meat will be increasing with rising income; wine consumption will remain constant. The need for bread, wheat, potatoes, pulses, and starchy food generally will diminish.

These conclusions are true for France as a whole, and the national plans take them into account in determining the long-term targets of agricultural production. They are also true for the other European countries of high standards of living. It may be noted that those foods whose consumption is assumed to increase are precisely the ones of greatest interest to the irrigated Mediterranean areas.

The new Lower Rhône-Languedoc plan for agriculture is resolutely directed towards meeting the needs of European consumers: growing fruits and vegetables, corn, and sorghum; intensive cultivation of alfalfa (in connection with producing meat) industrial crops (tomatoes for canning, sugar beets, durum wheat)—such are the principal targets. Although almost completely unexploited, France's export capacity of agricultural products is enormous. It has been said many times that France is a great agricultural nation and that it could be turned into Europe's granary or orchard and feed 70 million people. This may be so. But the facts show that up to now agriculture has played but a very small part in France's trade balance. As recently as 1952 France had an agricultural trade deficit. The situation has changed since then but agricultural exports are still low. On the average, French products represent hardly 4 per cent of Great Britain's large imports of fruit. Similarly, France's food exports to Germany are modest when measured in terms of the advantages offered by soil, climate, and proximity. In 1959 Germany imported more than seven times the value of dairy products from the Netherlands as from France, and her imports of fruits and vegetables from Italy were more than ten times those from France.

Through irrigation it is planned that agricultural income in the lower Rhône-Languedoc area will be gradually raised from the present level of about $75 million to approximately $150 million. With the necessary effort, we can see no difficulty in marketing these new supplies because European demand for high-grade products remains in excess of the supply potential; but the success of French producers in European markets will depend on better organization of production itself, a shortening of distribution channels, a bold and permanent export policy, and self-discipline in observing international quality standards.

THE EXECUTION OF THE AMBITIOUS DEVELOPMENT PLAN HAS BEEN assigned to the National Company for the Development of the Lower Rhône-Languedoc Area, which was organized in May, 1955. This organization is semigovernmental in form (société d'économie mixte). Many organizations of that type are functioning in France in a variety of fields. They have come to be a standard system when both public and private interests are involved in accomplishing a task which affects the public interest.

The many interests involved in the Lower Rhône-Languedoc project include central and local government, banks, industries, farmers and their organizations, power producers, and others. Their multiplicity is reflected in the ownership of the Company's capital stock, where, however, public bodies must always hold a majority interest.

The Company operates in the same way as a traditional private company. This gives it many advantages, especially in its day-to-day administrative work. The corporation is administered by a board of twelve. However, on account of the numerous tasks of public interest which are entrusted to it and because its financial resources are for the largest part public funds, it is subject to governmental control. The Board of

Administrators is assisted by the Director General of Rural Engineering of the Ministry of Agriculture who acts as Government Commissioner, a State Controller representing the Ministry of Finance, two general engineers acting as technical advisers and belonging, respectively, to the Ministry of Agriculture and the Ministry of Public Works. To complete the list of governmental officers or agencies, a special control branch has been set up and charged with ensuring that public funds are expended in conformity with government-approved projects and programs.

The organization of the Company, its role and working methods derive largely from the integrated character of the development project. From the beginning, it has been agreed that design, construction, and operation of the irrigation works should not be separated from economic development. Economic development is the actual aim of the enterprise; irrigation only the means. To ensure co-ordination and balance, it is essential that all actions, whatever their field of application—technical, economic, agricultural, psychological, financial—flow from the same authority and general policy.

Skipping over technical problems of design, construction, operation and maintenance of the irrigation works, it is sufficient here to say that the Company itself establishes the individual plans and designs and checks the work of the contractors chosen through competitive bids. For these purposes the Company has several technical offices and services with a present staff of 150. In the future, operating and maintaining such an irrigation network will require an additional 500 to 600 persons, even if electrical and hydraulic automation is pushed to the limits of current technology.

Scope of Company Operations. The government has granted the Company authority to undertake all construction and operations entailed by the irrigation projects, and the Company has been able to acquire, in the name of the State, all land necessary to any fixed structure included in the government-approved programs. Because all necessary construction has been qualified as public utility works, the Company has been granted the power to acquire land through dispossession proceedings. There is nothing really new in all this; similar patterns have been followed in other countries. More interesting is the way in which agricultural and economic development is being organized in terms of the soil, the peqple, and the means of production.

In Mediterranean areas the unbalanced climate, particularly the marked irregularity of rainfall, has contributed to land deterioration and has made irrigation a prerequisite of proper soil management. Consequently, soil tests are even more useful there than in most other places. The problems on a regional scale, as they present themselves to the engineer and on the scale of the individual holding, as they affect the farmer, call for a systematic progression of preliminary, intermediate scale, and detailed soil studies.

The aim of the preliminary study is to determine the irrigation sectors and to separate clearly the irrigable lands from those to be drained. It also makes it possible to compile an inventory of the different types of soils with a summary of their geographic location, and their characteristics. The intermediate scale study which includes the determination of the hydrodynamic and physiochemical soil characteristics, helps both in calculating the necessary irrigation and drainage works and establishing the crop programs. Detailed surveys complete and give precision to these first indications. A booklet containing the conclusions will be given each farmer so that every owner will know the nature of the soil on his farm, the appropriate crops, suitability for irrigation and appropriate form of irrigation (flooding or sprinkling).

The intermediate scale survey has been carried out so far on 215,000 acres, and detailed studies are being made at a yearly rate of 25,000 acres. It is the first time that a survey of this kind has been undertaken in France and indeed in Europe.

Among France's agricultural problems, the problem of scattered holdings may be considered among the most critical. In the Lower Rhône area, roughly one-third of the land consists of small holdings of less than 7.5 acres, one-third includes medium size properties ranging from 7.5 to 75 acres and the last one-third includes large estates of more than 75 acres. The cost of production is largely influenced by the size of the farm and even more by the location and shape of the different lots included in it. That is why it seems desirable to improve the farm structure by consolidating parcels of land and grouping farms into larger units, so as to ensure their economic viability by bringing cost and yield into balance and making efficient use of machinery. One effect of such improved land utilization might be to aggravate local problems of underemployment. Here the over-all program of regional development can help by opening new job opportunities through increased industrialization. For the area as a whole, however, the change from wine-growing to intensive irrigation agriculture should require a substantial increase in the number of farm workers; and the new farm processing industries also will need labor.

If consolidation is often necessary, so, on the other hand, the splitting up of large estates may be required as land reclamation and irrigation call for much higher investment and working capital than does dry farming. The owners may be capable of carrying on extensive cultivation but unable to face the new financial problems entailed by large-scale irrigation. Land consolidation, regrouping of small holdings, dividing of latifundia which are poorly adapted to intensive irrigated cultivation are the principal features of the land policy which the Company has been promoting for eighteen months.

A program of purchasing undercultivated land and reselling it to qualified farmers has been begun with the aim of putting together normal sized family farm units for intensive cultivation. An area amounting to 3,250 acres has already been acquired, constituting about 40 new farms. The Company is also concluding reclamation contracts with farmers who desire to improve their farm structure in order to be able to irrigate with the least delay. These contracts involve mostly land clearing, modernization of farm buildings, rural roads, drainage works, etc. Action along these lines will be intensified; twenty such contracts have been signed covering a total of 5,000 acres.

A series of farm bills voted during 1960 by Parliament will help to promote this policy. Farmers desiring to modernize their farming systems and increase their yields are eligible for special grants as well as short-, medium- and long-term loans. By contrast, owners of uncultivated lands, who take no action, are liable to dispossession.

Thus the land on which the new large-scale and scientific irrigation methods will be applied is itself on the way to being organized into a new pattern. Sprinkler irrigation is clearly tht best method. By reducing the volume of water needed and minimizing leaching and erosion, it is of particular advantage on the soils of the region, which are gravelly, porous, and deficient in humus. Also, sprinkling allows the farmer to apply irrigation any time he wants it instead of following the traditional but irrational practice of waiting his turn. The network of underground pipes required by sprinkler irrigation does not, like surface irrigation, freeze the present land pattern but leaves open the possibility of further land consolidation.

There is no doubt that the cost of this system is somewhat higher than that of surface irrigation. Exclusive of large structural works (intake works, canals, dams, and principal pumping plants), a surface irrigation network costs $200 per acre and a sprinkler network, under present methods, $270. However, by calculating the optimum diameters of pipes by means of electronic computers the investment can be lowered by 10 per cent. This would make the difference in capital cost between the two systems more bearable, especially in view of the operating savings that are possible because the sprinkler method requires less water per acre.

Still further reductions in investment are possible through replacement of steel and aluminum in both pipes and sprinklers by plastic materials that are much cheaper. Technology is moving rapidly in that field. Every type of sprinkling equipment is tested in the experiment station of the Company from both the technical and economic points of view.

The government and the Company are both conscious of their interest in improving agriculture through irrigation, but the farmers themselves do not always see the advantage so clearly. Most of them are wine-growers and lack experience in irrigation and other types of agriculture. They need information and advice. The Company has therefore set up a technical assistance service whose agents demonstrate new techniques, give advice, establish cultivation plans, help to launch farmers' associations, and offer other assistance. Winning the farmers' co-operation and good will is one of the most delicate problems the Company has to face. The psychology of the French farmer differs greatly from that of his American counterpart. Where the latter accepts willingly the extension agent's advice and is even eager for it, the former is reluctant. Success elsewhere, on the other hand, is certain to be emulated.

Technical and human reasons require friendly relations between farmers and agents. In our judgment it takes one agent for every 2,500 to 5,000 acres to accomplish this, with agricultural specialists in charge of the most difficult problems. One hundred field demonstrations per year are carried on, not at official experiment stations, but on ordinary farms whose owners are interested in the practice and in the promotion of efficient farm planning and irrigated cultivation.

But this will not suffice to speed up agricultural development. Governmental economic aid must reinforce technical action. Uprooting of vineyards and cultivation of new crops have been encouraged by grants amounting to $363 per acre. Since 1955, 42,000 acres of vineyards have been so eradicated, and rapid extension of the irrigation network should intensify this movement. Special loans that provide a four-year period without amortization are granted to fruit growers. Loans of the Land Credit Bank facilitate also the purchase of agricultural machinery and irrigation material and at the same time finance expenses of land improvement and house modernization. The interest rate ranges from 3 to 6.5 per cent for a repayment period from 30 to 9 years according to the type of investment. Special grower contracts guarantee an outlet for specific crops, either in fresh form or for canning.

The organization of marketing is another basic feature of

the region's economic development, though, of course, it must not be considered in the narrow framework of regional autonomy. Avoidance of useless competition between the different French provinces calls for national co-ordination that takes into account the characteristics and seasonality of the various regional crops, and the domestic and European demand for them according to seasons. A national organization capable of acting in that capacity has recently been formed. It brings together the National Credit Bank for Agriculture, the Foreign Trade National Bank, The Foreign Trade National Corporation, the National Corporation of Agricultural Cooperatives, and the National Central Association for Land Development. This organization is now working through offices and commercial agents already established in the principal cities and populous industrial areas of France and other parts of Europe.

Local farmers' associations are also being established, mostly in fruit growing, in the interest not only of better cultivation methods but of making farmers responsive to the consumer demand and to the need for observing European standards of quality production and marketing. Co-operatives or private corporations dealing especially with grading, packaging, and marketing complete the new set-up, whose possibilities should not be underestimated. Thanks to a similar policy, Italian growers have—through their FEDEXPORT organization—won a large share of the German market.

Marketing organization also implies the establishment of all sorts of agricultural processing industries, mainly the development of storage and quick-freezing plants and canneries for vegetables, fruits, and meats. Consideration is at present being given to such projects, co-operative as well as private. Annual production of fruits and vegetables could, within 25 years, rise to more than 1,200,000 tons. A number of well-known American and European food processors have found this prospect so attractive that they are planning to join the French efforts in the Lower Rhône project. The first step in that direction has been the construction of a large market and storage center near Nîmes. On a 60-acre site a fruit and vegetable market has already been built which will eventually be able to handle 300,000 tons of produce. Meanwhile other units are under construction: a cold storage plant with a capacity of over 400,000 cubic feet, a new cattle market, and a slaughter house with a capacity of 10,000 tons per year. In this connection it must be mentioned that Nîmes even now is the third largest cattle market in France, because of its location and its significance as an important railway junction.

Last but not least the living standards of the rural population must be raised, in step with rising farm output and income. Irrigation is incompatible with absentee ownership. Rural housing and community facilities have to be modernized, as in many other French regions. Closely allied is the need for new farm structures adapted to modern production and processing functions in line with the new directions in agriculture and the change to intensive cropping. Thus improvements must be made in the physical layout of villages (roads, public facilities, schools), quarters must be made available for co-operatives, farmers' unions, and similar activities, and the buildings themselves must be adapted to modern requirements. All this requires the co-operation and agreement of mayors and town councils. Long-term financial assistance is given by the central government; the municipalities and the farmers themselves have to share in the cost. To cut down expenses, it is preferable to concentrate construction activity in one village at a time or in a group of villages. A group of four villages has been by now completely studied and building has begun. The construction of 75 modern rural houses has been started in a new settlement area, while designs for ten more villages are under study. In the meantime 660 projects for the improvement of rural dwellings scattered all through the area have been studied, financed and undertaken.

About 140 persons are at present involved in projects of agricultural development but as soon as the irrigated area is sufficiently enlarged, something like 400 persons will be needed. Three different teams are at work: one is concerned with agricultural development problems in the economic sense, one with technical matters (construction and operation of the irrigation works), and the third team is responsible for general administration, personnel, legal and financial affairs. Co-ordination and stimulation are the job of the Director General who also handles the necessary contacts with the public authorities.

The Company is not in sole charge of all aspects of the region's economic development. Other agencies concern themselves with related problems that supplement the work of the Company, such as industrialization, tourism, and urban modernization. The Company co-operates fully with them. Indeed, all these activities spring from the same general policy and the Company is always a shareholder and member of the board of administrators of all these agencies. In this way overcentralization and overburdening of the Company can be avoided. Besides, it would be contrary to all French tradition for the government to allow a single authority to concentrate so many key jobs in its hands. We have not been able to duplicate in France, even on a smaller scale, the TVA experience.

Finances. The economic development project of the lower Rhône-Languedoc area is of interest to the nation's society as a whole and contributes to increasing the country's wealth. That is why the necessary investments are drawn from public funds, mainly through capital grants and long-term loans. The cost of the complete project is put at $200 million. Grants from the national government are generally provided for by enabling legislation covering three or four years. They are then confirmed in annual budgetary appropriations. Grants cover 60 or 75 per cent of the first cost depending upon the nature and characteristics of the complementary loans.

During the preliminary period (1956-59) government grants represented 60 per cent of total investment. Loans for the remaining 40 per cent were made directly by the Treasury following a decision by the Economic and Social Development Fund. They were to be amortized in 30 years and carried an interest rate of 1.25 per cent. The loans benefit from an initial 5-year grace period, so that repayment including interest works out to thirty annual payments, each 4 per cent of the total. During the years of deferred payment, the Company pays only interest.

From 1960 onwards, loans have been raised from other organizations. In 1960, the loans were obtained from the "Caisse des Dépots et Consignations," a public agency in charge of centralizing the Savings Banks resources and channeling them into public investments. The terms of these loans are less favorable. The amortization period is reduced to 25 years and the interest rate is raised to 5.25 per cent, but the contract retains the initial grace period of 5 years during which only interest is paid. A third method was begun in 1961. For the next 3 years, the required loans will be granted by the European Investment Bank—an organization created by the European Economic Community for sharing in the financing of economic development projects of European concern. The loan terms are even more stringent: the repayment period will not exceed 20 years and the interest rate is raised to 5.625 per cent. In both the 1960 and the current periods, the outright government grants represent 75 per cent of total investments in order to offset the higher loan charges.

Let us now briefly examine the Company's cost and income structure. Costs consist of (1) personnel and accessory expenses (salaries, travel, offices, cars, and social charges); (2) the cost of running the enterprise, involving (a) operation, maintenance and depreciation, and (b) electric power for pumping stations; and (3) the financial charges represented by the loan amortization. However, the financial agreement between the government and the Company provides that part of the latter charges may be refinanced through public funds as long as the utilization of the works has not reached a sufficient level. It is not hard to imagine how useful this relief is proving in the first years of the Company's life.

Revenue consists of (1) fees for various studies, and other services performed by the Company in connection with construction of the works, and for technical assistance to the farmers, and (2) of the payments for the water sold to the farmers, the towns, villages, and industries.

Payment of fees for studies will be spread over a period of about twenty years representing approximately the construction period. The fees are geared to the cost of the works and generally amount to between 5 and 8 per cent of those costs, varying with the degree of difficulty of the specific job. An additional 1.5 per cent is charged for technical assistance to the farmers and for agricultural as well as commercial plans for irrigation.

These receipts will constitute an appreciable part of the Company's working capital during its first years. But in the future they will decline to the vanishing point, while, at the same time, revenue from sale of water will increase until the time, still quite distant, when the irrigation works will be used at capacity. The bulk of the water will be used for agriculture; the remainder will go to towns and villages, and to industries. The selling price of irrigation water is based on its cost and the rates take the form of a classical two-part tariff. The price includes all costs: financial charges, personnel and administration, maintenance and depreciation, jointly estimated as a percentage of the cost of original construction (1 per cent for civil engineering, 2 per cent for pipes, 4 per cent for mechanical or electrical equipment) and cost of electric energy.

These costs are then divided into fixed and variable charges. In the two-part tariff, the fixed charge includes the fixed expenses: financial charges and maintenance and depreciation expenses. The variable charge includes the variable expenses of energy and personnel.

At present the average price of irrigation water is 5.6 cents per 1,000 gallons or $17.70 per acre-foot. This is made up of a fixed charge of around $10 per acre, and a variable charge of 3.1 cents per 1,000 gallons metered at the intake. The average cost of irrigation water works out to $24 per acre per year.

The price of water to towns, villages or industries is somewhat higher; it is presently set at 15.2 cents and 19.0 cents per 1,000 gallons, respectively. Should receipts exceed expenses, the profit has to be paid over to the government, as provided in the financial agreement signed by the Company. It may be considered as partial repayment of the grants which the government has made toward the financing of these investments. This procedure seems logical as the Company's public interest focus prevents it from making a profit.

LET ME CONCLUDE WITH A BRIEF SUMMARY. THE IRRIGATION works were started in 1957 at the end of a six-year period of technical and economic studies. The irrigation network is being developed at a yearly rate of 30,000 acres. At this rate the project will be completed around 1980. Since the start, investments have aggregated $70 million, mostly in permanent irrigation structures. That is why the irrigated area was only 3,750 acres last spring and is expected to reach no more than 10,000 acres during 1961. It is therefore too early to form a definite opinion of the future economic evolution of the lower Rhône-Languedoc area. But everything allows us to hope for a rapid and spectacular development both in agriculture and industry.

Present political events are causing many French living overseas to come and settle in the region, which is already called "French California." Free movement of persons inside the E.E.C. is expected also to bring Italians and Dutch. This fact alone, both irresistable and desirable, makes the development of France's underdeveloped areas necessary so as to lead to the rapid disappearance of what a young geographer has called the "French desert lands." Better tourist facilities in this region, one of the most beautiful in France, and industrial growth, of which there are already signs, will complement the agricultural development now under way.

Development of the Lower Rhône-Languedoc area will bring a balanced renewal of all economic and social activities, bring income stability for and better job opportunities to its inhabitants. This then represents a patient and methodical search for a greater blossoming of its human elements and a better use of its natural resources for the common good.

WHAT THE UNITED STATES CAN LEARN

■ STEFAN H. ROBOCK

PHILIPPE LAMOUR HAS MADE a stimulating and significant contribution in the field of regional resources development. His report on the experience in Southern France warrants serious attention from Americans despite substantial differences in the institutions, philosophies, and resources of France and the United States.

Among the differences in environment between the two countries, that of size—size of population, size of area, and size

■

STEFAN H. ROBOCK is director of international business studies and associate professor of international business at the Indiana University School of Business. Previously he was deputy director of the Area Development Division of the Committee for Economic Development (CED). After working as an economist for a number of government agencies, including the Antitrust Division of the U.S. Department of Justice and the National Resources Planning Board, he was chief economist of the Tennessee Valley Authority from 1949 to 1954. For the next two years he served with the United Nations Technical Assistance program as economic development adviser to Brazil, an experience that has been followed by several shorter assignments for the United Nations in South America. In 1959, he also served as a United Nations adviser on regional development programming in India. From 1956 to 1958 he was director of economic research for the Midwest Research Institute in Kansas City, Missouri. Mr. Robock is author of a number of books and professional articles in the field of economic development. He was born at Redgranite, Wisconsin, in 1915. He received his A.B. degree from the University of Wisconsin and his Ph.D. in economics from Harvard University.

of development projects—is one of the most important. In population, the United States is more than four times as big as France. And in area, even when Alaska and Hawaii are excluded, it is about fourteen times larger (3 million as compared to 212,000 square miles). Especially pertinent to the kind of regional development described by Mr. Lamour are the great size disparities in water resource projects. The Tennessee Valley watershed, only a medium-size resource development area for the United States, covers about 41,000 square miles as compared to the 1,000 square-mile area in the Lower Rhône-Languedoc project of Southern France.

The distribution of political power among levels of government is a second major difference. France has a highly centralized government, Mr. Lamour reminds us, and local governmental units have only limited autonomy. In contrast, the local and state governments in the United States have much greater relative power. This distribution of governmental authority has been an important factor in shaping our patterns of resource development.

Tendencies towards centralization and decentralization offer still another contrast. Mr. Lamour reports a noticeable tendency in France toward decentralization of economic, administrative, and political life. In the United States a reverse trend seems to exist. The federal government is playing an increasingly important and more centralized role in our economic life. At first glance one might suppose that these recent trends merely dilute the historical differences mentioned in the paragraph above. I do not think this is the case. The basic differences in governmental structure are so deep-rooted and pervasive that the recent reverse trends in each country simply constitute an added difference in environment.

A fourth difference relates to prevailing economic and political philosophies. In France important sectors of the economy have been nationalized. In the United States only a small part of our economic activity is directly operated by the government and in recent years popular attitudes have been strongly opposed to an increased role by government in the production sectors.

A final environmental difference worth noting here relates to historical patterns of growth. France has a large share of its population and economic activity in one metropolitan area. The United States has also been urbanizing, but due to the great size of the country a large number of nodal growth centers have been developing. The one-city dominance experienced in France, however, is quite similar to the geographical growth pattern encountered in most small countries.

But, in spite of all the differences, I am not suggesting that the United States has nothing to learn from the experience of Southern France. On the contrary, there is much in the record to be noted and pondered. Because of limitations of time as well as of professional competence, I have selected only five of the many interesting and significant features of the Lower Rhône and Languedoc project for discussion. My topics have not been ranked by importance, and they do not represent complete coverage of the experience that we might consider.

The most intriguing feature of Mr. Lamour's project in Southern France is the concept it represents. The program for this distressed area attempts to improve economic welfare through, to borrow a phrase from Harvey Perloff, "using natural resources that have yet to be fully tapped" and through creating new patterns of economic activity. In other words, an affirmative development program is being substituted for a disguised relief program that has been subsidizing economically obsolete agricultural patterns, namely the production of common wine.

The problem of economically lagging areas—also called distressed or underdeveloped areas—is a worldwide phenomenon.[1] Both capitalist and socialist countries have been experiencing uneven geographical development. Sometimes changes in technology and in consumption patterns cause economic activity in a previously thriving area to retrogress. In other situations, an area is considered to have economic problems because its growth rate is less than in other regions of the country.

[1] For a more extended discussion of this observation, see Stefan Robock, "Regional and National Economic Development in India," *Papers and Proceedings of the Regional Science Association*, Volume 6, 1960.

But whatever the reason for lagging development, you can be sure that strong local pressures will be generated to secure increased government aid for the area. When government assistance is secured, it is all too often directed toward subsidizing uneconomic activities or it is used in unsuccessful efforts to revive the past. And in innocent self-deception, the government officials and the local people invariably call such activities area development programs. If these comments appear to raise questions about certain currently popular legislative proposals in the United States, this is not coincidental. But I should add that I have encountered the same kind of self-deception in each of the technical assistance assignments I have undertaken in foreign countries.

As an economist, I would like to see the United States adopt the French approach of first investigating the resources and other development potentials of lagging areas and of then designing programs related to the realistic development potentials of these areas. I would go even further, however, and echo Harvey Perloff's conclusion in a recent study, that some areas will "have to face up to their very limited potentialities and to the urgent need for a high rate of out-migration."[2]

How refreshing it would be to take an affirmative rather than a remedial approach to the distressed area problem in the United States. And how exciting it is to think of the possibilities for accelerating national economic growth by substituting affirmative development programs for the fabulously expensive farm subsidy programs we have tolerated for so many years!

A second attractive feature of the French experience is the use of comprehensive economic pre-planning activities at the national level for regional resources development project. This pre-planning activity, according to Mr. Lamour, "had to take into account all the consequences and economic advantages of the plan's accomplishment."

Resources development projects which involve the invest-

[2] Harvey S. Perloff, "Lagging Sectors and Regions of the American Economy," *American Economic Review*, May, 1960.

ment of public funds by the federal government—and most such projects in the United States are of this type—should be planned at the national level, in relation to the national environment and justified on the basis of national benefits and priorities. Furthermore, the ultimate planning responsibility should not be in the hands of the agency which has a vested interest in the operation of the project.

Yet we have been reluctant to do such pre-planning at the national level. Even the highly successful Tennessee Valley Authority project did not have the benefit of comprehensive economic and social pre-planning. And I know of no other water resources project in the United States that has been based on the thorough and objective type of pre-planning that I have in mind.

The problem we face here is one imposed by semantics. In recent years, large segments of political and popular opinion have been persuaded that *planning by the federal government* is un-American, subversive, and even socialistic. At the same time, long-range planning has become a symbol of progress and of outstanding efficiency when used by private corporations. City planning is also completely acceptable and in great demand. Planning by state governments, though less prevalent, is generally tolerated. But federal planning—No. Those are evil words. In our present environment it is hard to believe that the United States had a National Resources Planning Board not so many years ago.

If there were in this country a central planning office—some kind of successor to the National Resources Planning Board—it would not have to go very deeply into the situations and alternative possibilities of the several regions. Both the detailed studies and the actual planning work could be done by regional groups. Instead, a technically qualified central group could review studies and plans from different areas from the standpoint of the nation-wide outlook and could, on occasion, supply information—special tabulations of Census data, for example—that communities or even regional establishments could not pull together for themselves. Above all, a central planning office would represent, as well as carry out, a general agreement that the national interest was the ultimate criterion for all federally supported programs of economic development.

Comprehensive federal planning offers great possibilities for achieving higher levels of efficiency and productivity in resources development. The pre-planning experience of France, therefore, deserves our attention. I should add, however, that our schizophrenic attitude toward planning is a critical issue for the United States reaching far beyond the resources development field.

Another feature of the Southern France project that might be more widely adopted in the United States is the emphasis on studies of future demand, particularly in the planning and implementation of our irrigation projects. Mr. Lamour reports that his agricultural program has been based upon future demand prospects for agricultural products in France and in possible export markets. All too frequently in the United States we have undertaken irrigation projects which have increased the supply of agricultural goods for which the demand was not necessarily increasing. He gives the United States credit for pioneering work in the study of such future demand patterns for agricultural products. But unfortunately in many cases this pioneering work has not been related to the planned output patterns and to benefit-cost analyses for resource development projects.

A fourth impressive aspect of the experience in Southern France is the attention given, in both planning and implementation, to complementary programs external to the resources development project. I refer here to the improvement in marketing institutions and practices as an auxiliary feature of the Southern France development.

Except for the Tennessee Valley Authority, our operating agencies for water resource projects have not done very well on this score. For example, our river navigation projects generally include only physical development activities for creating and maintaining a navigable channel. Essential complementary programs such as reserving waterside land for business firms that can make use of the river, creating industrial districts, building water freight terminals, and influencing changes in transportation rates are normally neglected because the operating agency does not consider such development activities as its responsibility. Yet, the physical improvement of resources is not meaningful in terms of long-run increases in income and employment unless complementary programs are adopted to make these resources *economically* useful.

A final feature of Mr. Lamour's project that deserves more attention in the United States is the use of the "mixed corporation." This type of institution is widely used throughout the world but virtually unknown in the United States.

The mixed corporation approach may have many advantages for the United States. It is a means of joining federal leadership with local responsibility and local financial participation. The mixed corporation also permits local private interests to influence the management practices and financial administration of resource development projects. The reasons why the mixed corporation has received so little attention in the United States are not all clear to me. This approach was widely used here during the last century. Louis Hartz, the historian, notes that the state of Pennsylvania participated as owner in some 150 mixed corporations.[3]

IN CONCLUSION I WOULD LIKE TO MAKE TWO GENERAL observations suggested by Mr. Lamour's account of his program in Southern France and its place in the broader plans of development for the whole country.

My first comment relates to the employment goals and employment effects of the Lower Rhône-Languedoc area. The justification for the project gives great emphasis to the employment needs there. Understandably, in view of the breadth of his presentation, Mr. Lamour does not go into details of how and to what extent the resource development project will create net new employment.

[3] Louis Hartz, *Economic Policy and Democratic Thought: Pennsylvania, 1776-1860* (Cambridge: Harvard University Press, 1948).

I am concerned about the employment question because many of the activities undertaken or in prospect, such as farm enlargement and the shift in farm crops, appear more likely to reduce than to increase employment. On the industrial side, I must also note that power-oriented industries, according to our experience in the United States, are not labor-intensive types of manufacturing. And the new employment accomplishments of rural industry programs in the United States have not been impressive.[4]

My second point relates to basic economic approach in regional development. After pointing out that 1 million new jobs are necessary in France before 1965 and 2 million before 1970, Mr. Lamour says that these jobs should not be located "in the part of the country that already enjoys a fuller development even though investments there would have the best multiplier effect. They must be distributed among all areas where efficacious economic development is possible." In carefully qualifying this general statement to show that he does not mean indiscriminate scattering of development, he mentions the desirability of creating a handful of metropolitan centers. This is an approach with which I thoroughly agree. Nevertheless, I believe that the general principle of diffusion is one that we Americans can well ponder deeply. The social or political "equity" justification for increased development in economic lagging areas is one that I have encountered widely in the United States and in foreign areas. In an extreme form, this position implies that economic development should be distributed evenly or more or less evenly across the geography of an entire country. In a more moderate form, it argues against urbanization trends and in favor of moving jobs to rural areas.

I have some serious questions about this kind of a justification for economic development programs. And I must confess that I am much impressed with the argument made by Professor Albert Hirschmann in a recent and stimulating study that

[4] Stefan H. Robock, "Rural Industries and Agricultural Development," *Journal of Farm Economics*, August 1952.

"interregional inequality of growth is an inevitable concomitant and condition of growth itself. Thus, in the geographical sense, growth is necessarily unbalance."[5]

I recognize that political and social factors must modify economic considerations and greatly influence public investment priorities. But I also feel that economic development technicians must understand that an almost inevitable feature of economic growth is a relative shift in income and employment from agricultural production and rural areas on the one hand to nonagricultural activities and urban centers on the other. If this point is recognized, resource planners will not limit their horizons to the development alternatives of either reviving employment in rural areas or allowing concentration to continue apace in the one major metropolitan center. A new alternative may be to stimulate development in several other urban centers with great development potential. It is encouraging to note that in France there are plans for creating such magnets to offset the pull of the Paris area.

Resource development projects are too frequently justified in terms of helping a geographical or political area rather than in terms of helping people. Yet shouldn't the true objective of economic development be to improve the economic welfare of people even if this involves out-migration?

To return to our main theme, the several features of the French experience that we have discussed—the substitution of economic development programs for relief programs, the preplanning activities at the national level, the planning emphasis on future demand for the output of resource projects, the recognition of related external factors and the programing of complementary activities and the use of the mixed corporation—are certainly important issues for our consideration in the United States. We are indebted to Mr. Lamour for making available to us the French experience with these techniques.

[5] Albert O. Hirschmann, *The Strategy of Economic Development* (New Haven: Yale University Press, 1958), pp. 183-84.

INDEX

F

G

O

P

Y

Z

For Product Safety Concerns and Information please contact our EU
representative GPSR@taylorandfrancis.com
Taylor & Francis Verlag GmbH, Kaufingerstraße 24, 80331 München, Germany

www.ingramcontent.com/pod-product-compliance
Ingram Content Group UK Ltd.
Pitfield, Milton Keynes, MK11 3LW, UK
UKHW022218190926
475702UK00003B/424
9781617260513